Advances in Bioceramics and Porous Ceramics V

Advances in Bioceramics and Porous Ceramics V

A Collection of Papers Presented at the 36th International Conference on Advanced Ceramics and Composites January 22–27, 2012 Daytona Beach, Florida

Edited by
Roger Narayan
Paolo Colombo

Volume Editors
Michael Halbig
Sanjay Mathur

A John Wiley & Sons, Inc., Publication

For general information on our other products and services or for technical support, please contact our Customer Care Department within the United States at (800) 762-2974, outside the United States at (317) 572-3993 or fax (317) 572-4002.

Wiley also publishes its books in a variety of electronic formats. Some content that appears in print may not be available in electronic formats. For more information about Wiley products, visit our web site at www.wiley.com.

Library of Congress Cataloging-in-Publication Data is available.

ISBN: 978-1-118-20596-9
ISSN: 0196-6219

Printed in the United States of America.

10 9 8 7 6 5 4 3 2 1

Contents

POROUS CERAMICS

Preface

This issue contains the proceedings of the "Next Generation Bioceramics"and "Porous Ceramics: Novel Developments and Applications" symposia of the 36th International Conference and Exposition on Advanced Ceramics and Composites (ICACC), which was held on January 22–27, 2012 in Daytona Beach, Florida, USA.

A rapidly developing area of ceramic science & technology involves research into ceramic materials that enhance the treatment of dental and medical conditions. Bioinspired and biomimetic ceramics, which imitate features of materials found in nature, have also stimulated significant interest in the ceramics community. The "Next Generation Bioceramics" symposium addressed several areas related to processing, characterization, modelling, and use of bioceramic materials, including biomineralization; advanced processing of bioceramic materials; bioinspired and biomimetic ceramic materials; self-assembled bioceramic materials; inorganic-organic composite materials; nanoscale bioceramic materials; in vitro and in vivo evaluation of bioceramic materials; mechanical properties of bioceramic materials; bioceramic materials for drug delivery; bioceramic materials for gene delivery; bioceramic materials for sensing; and bioceramic materials for dental applications. This symposium promoted lively discussions among various groups in the bioceramics community, including academic researchers, governmental researchers, and industrial researchers.

The "Porous Ceramics" symposium brought together engineers and scientists working in the area of highly porous ceramic materials, which contain a volume fraction of porosity typically higher than 70%, with pore size ranging from the nano- to the milli-meter scale. The presence of porosity is a key characteristic of these components, enabling their use in widely different and strategic areas such as environment, energy, defense, biomedicine, aeronautics, etc. The topics covered in the 3 days symposium ranged from innovations in processing methods and synthesis, structure and properties, modeling and novel characterization tools, mechanical behavior, micro- and meso-porous ceramics and ceramic membranes.

The quality of the oral and poster presentations and the good attendance were a testimony to the large interest that exists in the community, both academy and industry, for porous ceramics because of their peculiar characteristics and widespread applicability.

We would like to thank the staff at The American Ceramic Society, particularly Greg Geiger, Mark Mecklenborg, Marilyn Stoltz, and Marcia Stout for making this volume possible. We would also like to thank Anita Lekhwani and her colleagues at John Wiley & Sons for their efforts in support of this volume. In addition, we would like to acknowledge the contributors and reviewers, without whom this volume would have not been possible. We also thank the officers of the Engineering Ceramics Division of The American Ceramic Society, including Michael Halbig, Sanjay Mathur, Tatsuki Ohji, Dileep Singh, Mrityunjay Singh, Sujanto Widjaja, and the 2012 Program Chair,

Prof. Sanjay Mathur, for their tireless efforts. We hope that this volume becomes a useful resource for academic and industrial efforts involving porous ceramic materials and bioceramic materials. Finally, we anticipate that this volume contributes to advances in ceramic science & technology and signifies the leadership of The American Ceramic Society in these emerging areas.

ROGER NARAYAN,
University of North Carolina and North Carolina State University

PAOLO COLOMBO,
Università di Padova (Italy) and The Pennsylvania State University

Introduction

This issue of the Ceramic Engineering and Science Proceedings (CESP) is one of nine issues that has been published based on content presented during the 36th International Conference on Advanced Ceramics and Composites (ICACC), held January 22–27, 2012 in Daytona Beach, Florida. ICACC is the most prominent international meeting in the area of advanced structural, functional, and nanoscopic ceramics, composites, and other emerging ceramic materials and technologies. This prestigious conference has been organized by The American Ceramic Society's (ACerS) Engineering Ceramics Division (ECD) since 1977.

The 36th ICACC hosted more than 1,000 attendees from 38 countries and had over 780 presentations. The topics ranged from ceramic nanomaterials to structural reliability of ceramic components which demonstrated the linkage between materials science developments at the atomic level and macro level structural applications. Papers addressed material, model, and component development and investigated the interrelations between the processing, properties, and microstructure of ceramic materials.

The conference was organized into the following symposia and focused sessions:

Symposium 1	Mechanical Behavior and Performance of Ceramics and Composites
Symposium 2	Advanced Ceramic Coatings for Structural, Environmental, and Functional Applications
Symposium 3	9th International Symposium on Solid Oxide Fuel Cells (SOFC): Materials, Science, and Technology
Symposium 4	Armor Ceramics
Symposium 5	Next Generation Bioceramics

Symposium 6	International Symposium on Ceramics for Electric Energy Generation, Storage, and Distribution
Symposium 7	6th International Symposium on Nanostructured Materials and Nanocomposites: Development and Applications
Symposium 8	6th International Symposium on Advanced Processing & Manufacturing Technologies (APMT) for Structural & Multifunctional Materials and Systems
Symposium 9	Porous Ceramics: Novel Developments and Applications
Symposium 10	Thermal Management Materials and Technologies
Symposium 11	Nanomaterials for Sensing Applications: From Fundamentals to Device Integration
Symposium 12	Materials for Extreme Environments: Ultrahigh Temperature Ceramics (UHTCs) and Nanolaminated Ternary Carbides and Nitrides (MAX Phases)
Symposium 13	Advanced Ceramics and Composites for Nuclear Applications
Symposium 14	Advanced Materials and Technologies for Rechargeable Batteries
Focused Session 1	Geopolymers, Inorganic Polymers, Hybrid Organic-Inorganic Polymer Materials
Focused Session 2	Computational Design, Modeling, Simulation and Characterization of Ceramics and Composites
Focused Session 3	Next Generation Technologies for Innovative Surface Coatings
Focused Session 4	Advanced (Ceramic) Materials and Processing for Photonics and Energy
Special Session	European Union – USA Engineering Ceramics Summit
Special Session	Global Young Investigators Forum

The proceedings papers from this conference will appear in nine issues of the 2012 Ceramic Engineering & Science Proceedings (CESP); Volume 33, Issues 2-10, 2012 as listed below.

- Mechanical Properties and Performance of Engineering Ceramics and Composites VII, CESP Volume 33, Issue 2 (includes papers from Symposium 1)
- Advanced Ceramic Coatings and Materials for Extreme Environments II, CESP Volume 33, Issue 3 (includes papers from Symposia 2 and 12 and Focused Session 3)
- Advances in Solid Oxide Fuel Cells VIII, CESP Volume 33, Issue 4 (includes papers from Symposium 3)
- Advances in Ceramic Armor VIII, CESP Volume 33, Issue 5 (includes papers from Symposium 4)

- Advances in Bioceramics and Porous Ceramics V, CESP Volume 33, Issue 6 (includes papers from Symposia 5 and 9)
- Nanostructured Materials and Nanotechnology VI, CESP Volume 33, Issue 7 (includes papers from Symposium 7)
- Advanced Processing and Manufacturing Technologies for Structural and Multifunctional Materials VI, CESP Volume 33, Issue 8 (includes papers from Symposium 8)
- Ceramic Materials for Energy Applications II, CESP Volume 33, Issue 9 (includes papers from Symposia 6, 13, and 14)
- Developments in Strategic Materials and Computational Design III, CESP Volume 33, Issue 10 (includes papers from Symposium 10 and from Focused Sessions 1, 2, and 4)

The organization of the Daytona Beach meeting and the publication of these proceedings were possible thanks to the professional staff of ACerS and the tireless dedication of many ECD members. We would especially like to express our sincere thanks to the symposia organizers, session chairs, presenters and conference attendees, for their efforts and enthusiastic participation in the vibrant and cutting-edge conference.

ACerS and the ECD invite you to attend the 37th International Conference on Advanced Ceramics and Composites (http://www.ceramics.org/daytona2013) January 27 to February 1, 2013 in Daytona Beach, Florida.

MICHAEL HALBIG AND SANJAY MATHUR
Volume Editors
July 2012

Bioceramics

EFFECT OF PRECURSOR SOLUBILITY ON THE MECHANICAL STRENGTH OF HAP BLOCK

Nurazreena Ahmad[1], Kanji Tsuru[1], Melvin L. Munar[1], Shigeki Matsuya[2] and Kunio Ishikawa[1]

[1] Department of Biomaterials, Faculty of Dental Science, Kyushu University, JAPAN
[2] Sections of Bioengineering, Department of Dental Engineering, Fukuoka Dental College, JAPAN

ABSTRACT

The effect of the solubility of the precursors, alpha tricalcium phosphate (α-TCP) and beta tricalcium phosphate (β-TCP) on the mechanical property of hydroxyapatite (HAp) bone substitute was investigated. Uniaxially pressed compacts starting from these precursors were treated hydrothermally with 1 mol/L of ammonia solution at 200°C for various durations. XRD analysis revealed that α-TCP took 3 hours whereas β-TCP took 240 hours for complete transformation to HAp. The porosity of HAp block obtained from β-TCP was found to be lower than that of HAp block from α-TCP. Diametral tensile strength of HAp block from β-TCP showed a significantly higher value than that of HAp block obtained from α-TCP. It is therefore concluded that solubility of precursor affects the mechanical strength of the HAp block.

INTRODUCTION

Hydroxyapatite (HAp; $Ca_{10}(PO_4)_6(OH)_2$) have been used clinically to reconstruct bone defects since it shows excellent tissue compatibility and osteoconductivity[1].

Sintering is the most common method in fabricating HAp bone substitute. However, crystallinity of sintered HAp is extremely high[1]. HAp in bone shows low crystallinity. Furthermore, it is known that HAp with lower crystallinity shows better osteoconductivity[2]. HAp block with low crystallinity can be fabricated through phase transformation using thermodynamically unstable precursor based on dissolution-precipitation reaction by hydrothermal treatment[3]. For the dissolution-precipitation reaction, one requirement for the precursor is to have a moderate solubility when compared to the final block product. Due to the instability of the precursor, it is dissolved in aqueous solution and supply ions that are required for the precipitation of the final product[3]. Precursors with high solubility will dissolve faster while precursor with low solubility will dissolve slower. High solubility of precursor will cause rapid precipitation of crystals and may result in a mismatch crystal arrangement and/or more porosity. In contrast, precursor with low solubility will dissolve slower and thus the precipitated crystals may be arranged closely with less porosity.

The purpose of this study is to evaluate the effect of precursor's solubility on the mechanical property on the HAp block through dissolution-precipitation reaction. For the precursor selection, alpha-tricalcium phosphate (α-TCP; α-$Ca_3(PO_4)_2$) and beta-tricalcium phosphate (β-TCP; β-$Ca_3(PO_4)_2$) were employed as the precursors due to the same chemical composition but of different solubility. α-TCP has been reported to have higher solubility in aqueous solution when compared to β-TCP[4].

MATERIALS AND METHODS

Preparation of α-TCP and β-TCP blocks

Commercially obtained calcium hydrogen phosphate dihydrate (DCPD; $CaHPO_4 \cdot 2H_2O$, Wako, Osaka, Japan) and calcium carbonate (Calcite; $CaCO_3$, Wako) were mixed with 100 ml of distilled water in 2:1 molar ratio so that the Ca/P molar ratio would be 1.5. The mixing was done using rotary pestle and mortar for 24 hours. After that, the mixture was dried in a 60°C oven (DO-450FA, As One Corp., Osaka, Japan) for 24 hours and then crushed into powder form. The powder was then pressed uniaxially with an oil press machine (Riken Power, Riken Seiki, Tokyo, Japan) at 5 MPa pressure. The blocks were then sintered at 800°C in a furnace (SBV-1515D, Motoyama, Osaka, Japan) for 12 hours to allow homogenous composition of calcium and phosphate ions. The sintered blocks were then again crushed into powder form followed by grinding in a planetary micro mill (Fritsch, Idar-Oberstein, W. Germany). After grinding, the powder was pressed again with an oil press machine at 5 MPa pressure to obtain a block-type specimen 6 mm in diameter and 3 mm in thickness.

To obtain α-TCP block, the blocks were sintered at 1250°C in a furnace and kept for 4 hours then quenched to room temperature to avoid α to β phase transformation during cooling. To obtain β-TCP block, the blocks were sintered at 1100°C and kept 4 for hours followed by quenching to room temperature. The heating rate used was 10°C/min.

Hydrothermal Treatment

The α-TCP and β-TCP blocks were subjected to hydrothermal treatment with 1 mol/L of ammonia (NH₃; Kanto Chemical Co., INC., Tokyo, Japan) solution. One α- or β-TCP block were immersed in 30 ml of NH₃ solutions and placed in hydrothermal vessel (Shikoku Rika Co. Ltd, Kochi, Japan) which consisted of a Teflon® inner vessels with stainless steel jacket. The hydrothermal vessels were kept at 200°C in a drying oven for various durations. After the hydrothermal reaction, the blocks were removed from the solution, washed 20 times with distilled water to remove NH₃ solution and dried at 60°C for 24 hours.

X-Ray Diffraction Analysis

For compositional analysis, the specimens were characterized by powder X-ray diffraction, XRD (D8 Advance A25 Bruker AXS GmbH, Karlsruhe, Germany) using counter-monochromatic CuKα radiation generated at 40 kV and 40 mA.

Porosity Measurement

The bulk density of the α-TCP and β-TCP blocks before and after hydrothermal treatment was calculated based on the weight and volume. To measure the porosity, the relative density of the blocks was calculated based on their bulk densities and theoretical densities of α-TCP (2.86 g/cm³), β-TCP (3.07 g/cm³) and HAp (3.14 g/cm³) as shown in equation 1. The total porosity of the block was then defined as in equation 2. Total porosity was the mean value of at least 6 blocks.

$$\text{Relative density (\%)} = \text{Bulk density / Theoretical Density} \times 100\% \qquad (1)$$

$$\text{Total porosity (\%)} = 100\ (\%) - \text{Relative density (\%)} \qquad (2)$$

Mechanical Strength Measurement

Mechanical strength of the specimens was evaluated in terms of diametral tensile strength (DTS). The specimens were tested using a universal testing machine (AGS-J, Shimadzu Corporation, Kyoto, Japan) by crushing it at a crosshead speed of 1 mm/min. The DTS values used were average of at least 8 specimens. For statistical analysis, one-way factorial ANOVA and Fisher's PLSD method as a post-hoc test were performed using KaleidaGraph 4.0.

Scanning Electron Microscope

Morphology of the fractured surface was observed using scanning electron microscope, SEM (S-3400N, Hitachi High Technologies Co., Tokyo, Japan) at an accelerating voltage of 10 kV. The fractured surface was coated with gold prior to SEM observation.

RESULTS AND DISCUSSION

Porosity of both α-TCP and β-TCP blocks prepared by sintering methods was approximately 10%. Fortunately, sintering temperature and resulting phase caused almost no effect on the porosity of the precursor. After hydrothermal treatment at 200°C in 1 mol/L NH_3 solution, both α-TCP and β-TCP blocks were transformed to HAp blocks. However, the length of time required for the complete transformation to apatite was different up to the precursors. In other words, the β-TCP block needed longer time, 240 hours, to obtain single phase HAp. In contrast, α-TCP required only 3 hours to obtain single phase HAp.

SEM observation demonstrated different feature after the hydrothermal treatment. Before hydrothermal treatment, both α-TCP and β-TCP block showed smooth surface typical for sintering process. However, needle-like crystals interlocked each other after hydrothermal treatment. It should be noted that the crystals shape and arrangement were different based on the precursors i.e. α-TCP or β-TCP as shown in figure 1. On the surface of HAp block fabricated from α-TCP, many small crystals were formed while the crystals formed were large in the case of β-TCP. It is observed that the HAp crystals from α-TCP were loosely attached with obvious porosity. In contrast, the HAp crystals from β-TCP were arranged closely to each other with no obvious porosity. The difference in crystal morphology is thought to be caused by the effect of precursor solubility. During hydrothermal treatment, dissolution-precipitation reaction would occur and this leads the compositional transformation from TCP to HAp without changing the macroscopic structure. However, rate of dissolution is thought to affect the precipitation process. When α-TCP is used as the precursor, shorter time is required for the solution to reach critical supersaturations with respect to apatite when compared to that of β-TCP since the solubility of α-TCP is higher than β-TCP. Thus, structural mismatch due to the weak interactions between the nuclei might inhibit crystals arrangements and less ordered crystal structures were formed. For β-TCP as the precursor, a more ordered crystals structures were formed since the low solubility creates slower time to reach supersaturations with respect to apatite[5]. This structural mismatch results in the higher amount of porosity in HAp block from α-TCP compared to HAp block from β-TCP.

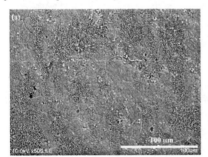

Figure 1. SEM on the surface of (a) HAp from α-TCP (b) HAp from β-TCP

The porosities of HAp block was 17±0.9% and 14±0.6% when α-TCP and β-TCP were used as precursors, respectively ($p < 0.05$).

The DTS value of HAp block obtained from β-TCP and α-TCP based on hydrothermal treatment in NH_3 solution at 200°C for 240 hours and 3 hours were 17±2.8 MPa and 11±1.15 MPa, respectively ($p < 0.05$). In other words, HAp block fabricated from□β-TCP showed higher DTS value when compared to the value obtained from HAp block fabricated from α-TCP. The high DTS value of HAp block fabricated from β-TCP is thought to be caused by the less porosity. The relationship between the mechanical strength and porosity is well expressed by equation 3.

$$S = S_0 \exp(-bP) \tag{3}$$

where S is the observed mechanical strength of the porous material, S_0 is the ideal mechanical strength when there is no porosity, P is the porosity, and b is the empirical constant. In this equation, it is clear that less pore is the key for higher mechanical strength[6]. Further study is awaited based on the initial findings obtained in this preliminary study.

CONCLUSION

In this study, it was found that β-TCP which has low solubility when compared to α-TCP transformed to HAp with higher mechanical strength. Therefore it is concluded that the selection of precursors with suitable solubility is an important factor in determining the final mechanical property of the HAp bone substitute formed based on dissolution-precipitation reaction.

ACKNOWLEDGEMENTS

This study was supported in part by a grant-in-aid for scientific research from the Ministry of Education, Culture, Sports, Science, and Technology of Japan, as well as by the scholarship from the Ministry of Higher Education Malaysia and University Science Malaysia.

REFERENCES
[1]W. Suchanek and M. Yoshimura, Processing and Properties of Hydroxyapatite Based Biomaterials for use as Hard Tissue Replacement Implants, *J. Mater. Res.*, **13**, 94-117 (1998).
[2]H. Oonishi, L. L. Hench, J. Wilson, F. Sugihara, E. Tsuji, S. Kushitani, and H. Iwaki, Comparative Bone Ingrowth Behavior in Granules of Bioceramic Materials of Various Sizes, *J. Biomed. Mater. Res.*, **44**, 31-43 (1999).
[3]K. Ishikawa, Bone Substitute Based on Dissolution-Precipitation Reactions, *Materials*, **3**, 1138-55 (2010).
[4]P. Ducheyne, S. Radin and L. King, The Effect of Calcium Phosphate Ceramic Compositin and Structure on *In Vitro* Behaviour I: Dissolution, *J. Biomed. Mater. Res.*, **27**, 25-34 (1993).
[5]L. Wang and G. H. Nancollas, Calcium Orthophosphate: Crystallization and Dissolution, *Chem. Rev.*, **108**, 4628-4669 (2008).
[6]K. Ishikawa and K. Asaoka, Estimation of Ideal Mechanical Strength and Critical Porosity of Calcium Phosphate Cement, *J. Biomed. Mater. Res.*, **29**, 1537-1543 (1995).

CARBONATE APATITE FORMATION DURING THE SETTING REACTION OF APATITE CEMENT

Arief Cahyanto, Kanji Tsuru, and Kunio Ishikawa
Department of Biomaterials, Faculty of Dental Science, Kyushu University, Fukuoka, JAPAN

ABSTRACT
Replacement of apatite cement (AC) to bone is still controversial issue. To understand factor that could affect the replacement of AC to bone, AC consisting of an equimolar mixture of tetracalcium phosphate (TTCP; $Ca_4(PO_4)_2O$) and dicalcium phosphate anhydrous (DCPA; $CaHPO_4$) was allowed to set at 37°C and 100% relative humidity under 5% CO_2 or N_2. Carbonate apatite (CO_3Ap) was formed when AC was allowed to set under 5% CO_2. The amount of CO_3 decreased gradually as the depth from the surface increased. The CO_3Ap was the B-type CO_3Ap in which CO_3^{2-} was replaced with PO_4^{3-} and the CO_3Ap found in bone. Larger amount of TTCP remain unreacted when the AC was allowed to set under N_2 whereas smaller amount of TTCP remain unreacted when the AC was allowed to set under CO_2. This may be caused by the larger Ca/P molar ratio of CO_3Ap. Formation of CO_3Ap and/or small unreacted TTCP are thought to be key factors for the replacement of AC to bone.

INTRODUCTION
Apatite cement (AC) that set to form apatite was a breakthrough for the reconstruction of bone defect.[1-3] AC shows excellent tissue response and good osteoconductivity. It should be noted that replacement of AC is still in chaos. Some reported that AC replaced with bone with time.[4-6] On the other hand, some reported that AC would not be replaced to bone at all.[7-9] These different observations concerning the replacement of AC to bone is thought to be caused by the different clinical cases. Ishikawa et al. reported that CO_3Ap was found on the surface of set AC consisting of an equimolar mixture of tetracalcium phosphate (TTCP; $Ca_4(PO_4)_2O$) and dicalcium phosphate anhydrous (DCPA; $CaHPO_4$) even though no carbonate source was present in the AC.[10] On the other hand, CO_3Ap was reported to be replaced to bone. Therefore, formation of CO_3Ap maybe the key factor for the replacement of AC to bone. However, no detailed mechanism has been clarified why CO_3Ap was found.

Obviously, carbonate ion could be supplied from atmosphere and/or the body fluid. In the present study, therefore, effect of setting atmosphere on the composition of set AC was studied by setting TTCP-DCPA based AC under 5% CO_2 or 100% N_2 atmosphere at 37°C and 100% relative humidity.

MATERIALS AND METHODS

Preparation of AC Powder
TTCP powder obtained commercially (Taihei Chemicals Inc, Osaka, Japan) was used without further modification. In the case of DCPA, particle size of commercially obtained DCPA powder (J.T. Baker Chemical Co., NJ, USA) was reduced to 1.4 μm by grinding the powder in a planetary micro mill (Fritsch 8 6580, Idar-Oberstein, Germany) with 95% ethanol for 24 hours followed by drying overnight at 40°C in a dry-heat oven (DO-300A, AS ONE, Japan).

Specimen Preparation
The powder phase was mixed with distilled water at a liquid to powder ratio of 0.3 (L/P ratio). The paste obtained was packed into a polycarbonate mold consisting of a pile of polycarbonate sheet

7

with 10 mm in diameter and 0.2 mm in height. After packing, one end of the mold was covered with a glass slide and clamped with a metal clip. The mold with the cement paste was then placed into an incubator under 5% CO_2 kept at 37°C and 100% of relative humidity for 24 hours. For control, the mold with cement paste was also placed in 100% under N_2 atmosphere at 37°C and 100% of relative humidity for 24 hours. Then, the set AC was immersed in the 99% ethanol for 3 min to arrest the reaction, and dried in an oven at 60°C for 2 hours. After removing the surface polycarbonate sheet, the surface specimen layer was scraped and subjected to the compositional characterization. The specimen collected from surface layer is denoted as "layer 1" and the specimen collected from the fifth layer from surface is denoted as "layer 5".

Characterization of Set Apatite Cement

The set AC was characterized by means of X-ray diffraction (XRD: D8 Advance, Bruker AXS GmbH., Karlsruhe, Germany) and Fourier transform infrared spectroscopy (FT-IR: FT/IR-6200, JASCO, Tokyo, Japan).

RESULTS AND DISCUSSION

XRD analysis of the specimen collected from each layer of set AC after being allowed to set for 24 hours under 5% CO_2 atmosphere and 100% N_2 atmosphere showed that transformation of AC to apatitic mineral was observed after 24 hours regardless of the atmospheric condition or the layer of specimens although some TTCP remain unreacted in the set AC. It should be noted that the amount of unreacted TTCP was different based on the atmosphere. In other words, smaller amount of TTCP remain unreacted when AC was allowed to set under 5% CO_2 atmosphere whereas larger amount of TTCP remain unreacted when AC was allowed to set under 100% N_2 atmosphere.

FT-IR spectra of powder specimen collected from each layer of AC before and after 24 hours under 5% CO_2 atmosphere or 100% N_2 atmosphere also demonstrated that set AC became apatitic mineral regardless of the atmospheric condition or the layer of specimen. In other words, peaks assigned to phosphate of apatite were detected at 560-970 and 1010-1100 cm^{-1}. For AC that set in 5% CO_2 atmosphere, some additional peaks were found at around 873, 1420 and 1470 cm^{-1} which were assigned to carbonate. These peaks are attributed to CO_3^{2-} ions in B-type CO_3Ap in which CO_3^{2-} is replaced with PO_4^{3-}.[11-13] The intensity of the carbonate peaks was larger when the specimen were taken from the surface layer when compared to the specimen taken deep inside the mold. These results are reasonable since CO_2 in the 5% CO_2 may penetrate to the AC paste from the exposed surface.

It is well known that setting reaction of the AC is based on dissolution-precipitation process and interlocking of precipitated apatite crystals.[14-15] As the dissolution process, both TTCP and DCPA dissolves to supply Ca^{2+} and PO_4^{3-} as shown in equation (1) and (2).

$$Ca_4(PO_4)_2O + H_2O \rightarrow 4Ca^{2+} + 2PO_4^{3-} + 2OH^- \qquad (1)$$

$$CaHPO_4 \rightarrow Ca^{2+} + H^+ + PO_4^{3-} \qquad (2)$$

In the absence of carbonate ion, the solution containing Ca^{2+} and PO_4^{3-} would be supersaturated with respect to HAp, and thus HAp crystals would be precipitated as shown in equation (3).

$$10Ca^{2+} + 6PO_4^{3-} + 2OH^- \rightarrow Ca_{10}(PO_4)_6(OH)_2 \qquad (3)$$

The precipitated HAp crystals interlock each other to form set HAp. On the other hand, the solution would contain not only Ca^{2+} and PO_4^{3-} but also CO_3^{2-} when the paste is exposed to CO_2. The solution would be supersaturated with respect to CO_3Ap, and thus CO_3Ap crystals would be precipitated as shown in equation (4).

$$(10\text{-a})Ca^{2+} + (6\text{-b})PO_4^{3-} + cCO_3^{2-} + (2\text{-d})OH^- \rightarrow Ca_{10\text{-a}}(PO_4)_{6\text{-b}}(CO_3)_c(OH)_{2\text{-d}} \quad (4)$$

The precipitated CO_3Ap crystals interlock each other to form set CO_3Ap.

At clinical situation, time and/or the area of AC exposed to CO_2 and/or CO_3^{2-} may be different. Therefore, replacement of AC to bone may be different up to the reports. Further research is awaited based on the initial findings found in this study.

CONCLUSION

Incorporation of CO_3^{2-} to AC during the setting reaction of AC may be the key issue of formation of CO_3Ap. The different CO_3^{2-} ions incorporation may cause a different behavior of AC to bone replacement.

ACKNOWLEDGEMENTS

The authors are grateful to acknowledge Kyushu University, Japan for conference financial support and Ministry of Education and Culture, Directorate of Higher Education (DIKTI), Republic of Indonesia for doctoral scholarship. The Departments of Dental Material Science and Technology, Faculty of Dentistry, Universitas Padjadjaran, Indonesia is acknowledged for supporting Arief Cahyanto in the framework of doctoral study.

REFERENCES
[1] W. E. Brown and L. C. Chow, Combinations Of Sparingly Soluble Calcium Phosphates In Slurries And Paste As Mineralizers And Cements, *U.S. Patent* No. 4,612,053 (1986).
[2] W. E. Brown and L. C. Chow, A New Calcium Phosphate, Water-Setting Cement, In Cements Research Progress, P. W. Brown (ed.), *Am. Ceram. Soc.*, Westerville, OH, 351–379 (1986).
[3] L. C. Chow and S. Takagi, A Natural Bone Cement—A Laboratory Novelty Led to the Development of Revolutionary New Biomaterials. *J. Res. Natl. Inst. Stand. Technol.*, **106**, 1029–1033 (2001)
[4] R. Mai, A. Reinstorf, E. Pilling, M. Hlawitschka, R. Jung, M. Gelinsky, M. Schneider, R. Loukota, W. Pompe, U. Eckelt, B. Stadlinger, Histologic Study of Incorporation and Resorption of A Bone Cement-Collagen Composite: An In Vivo Study in The Minipig, *Oral Surg. Oral Med. Oral Pathol. Oral Radiol. Endod.*, **105**, e9-e14 (2008).
[5] B. R. Constantz, I. C. Ison, M. T. Fulmer, R. D. Poser, S. T. Smith, M. Van Wagoner, J. Ross, S. A. Goldstein, J. B. Jupiter, and D. I. Rosenthal. Skeletal Repair by In Situ Formation of The Mineral Phase of Bone. *Science*, **267**, 1796- 1799 (1995).
[6] H. Yuan, Y. Li, J. D. de Bruijn, K. de Groot, X. Zhang, Tissue Responses of Calcium Phosphate Cement: A Study in Dogs, *Biomaterials*, **21**, 1283-1290 (2000).
[7] D. C. J. Cancian, E. H. Vieira, R. A. C. Marcantonio, I. R. Garcia Jr. Utilization of Autogenous Bone, Bioactive Glasses, and Calcium Phosphate Cement in Surgical Mandibular Bone Defects in Cebus apella Monkeys. *Int. J. Oral Max. Impl.*, **19**, 73-79 (2004).
[8] A. K. Gosain, P. A. Riordan, L. Song, M. Amarante, B. Kalantarian, P. G. Nagy, C. R. Wilson, J. M. Toth, B. L. McIntyre. A 1-Year Study of Osteoinduction in Hydroxyapatite-Derived Biomaterials in an Adult Sheep Model: Part II. Bioengineering Implants to Optimize Bone Replacement in Reconstruction of Cranial Defects. *Plast. Reconstr. Surg.*, **114**, 1155-1163 (2004)
[9] A. Maniker, S. Cantrell, C. Vaicys. Failure of Hydroxyapatite Cement to Set in Repair of a Cranial Defect: Case Report. *Neurosurgery*, **43**, 9534 (1998).
[10] Y. Miyamoto, K. Ishikawa, M. Takechi, T. Toh, Y. Yoshida, M. Nagayama, M. Kon, K. Asaoka. Tissue response to fast-setting calcium phosphate cement in bone. *J. Biomed. Mater. Res.*, **37**, 457-

464 (1997).

[11]R. Z. Le Geros, Apatites in Biological System, *Crystal Growth Charact.*, **4**, 1-45 (1981)

[12]J. P. Lafon, E. Champion and D. Bernache-Assollant. Processing of AB-type Carbonated Hydroxyapatite $Ca_{10-x}(PO_4)_{6-x}(CO_3)_x (OH)_{2-x-2y}(CO_3)_y$ Ceramics with Controlled Composition. *J. Eur. Ceram. Soc.*, **28**, 139-147 (2008).

[13]C. Rey, B. Collins, T. Goehl, I. R. Dickson, and M.J.Glimcher. The Carbonate Environment in Bone Mineral: A Resolution-Enhanced Fourier Transform Infrared Spectroscopy Study. *Calcif. Tissue Int.*, **45**, 157-164 (1989)

[14]K. Ishikawa, Bone Substitute Fabrication Based on Dissolution-Precipitation Reactions, *Materials*, **3**, 1138-1155 (2010).

[15]K. Ishikawa, S. Matsuya, Bioceramics: In Comprehensive Structural Integrity, I. Milne, R.O. Ritchie, B. Karihaloo, Eds., *Elsevier*, Oxford, UK, **9**, 169-214 (2003).

IN VITRO EVALUATION OF SILICATE AND BORATE BIOACTIVE GLASS SCAFFOLDS PREPARED BY ROBOCASTING OF ORGANIC-BASED SUSPENSIONS

Aylin M. Deliormanlı, Mohamed N. Rahaman
Missouri University of Science and Technology, Department of Materials Science and Engineering, and Center for Bone and Tissue Repair and Regeneration, Rolla, MO 65409, USA

ABSTRACT

Porous three-dimensional scaffolds of silicate (13-93) and borate (13-93B3) bioactive glass were prepared by robocasting and evaluated in vitro for potential application in bone repair. Organic-based suspensions were developed to limit degradation of the bioactive glass particles, and deposited layer-by-layer to form constructs with a grid-like microstructure. After binder burnout, the constructs were sintered for 1 hour at 690 °C (13-93 glass) or 560 °C (13-93B3 glass) to produce scaffolds (porosity ≈ 50%) consisting of dense glass struts (400 μm in diameter) and interconnected pores of width 300 μm. The mechanical response in compression and the conversion of the scaffolds to hydroxyapatite (HA) were studied as a function of immersion time in a simulated body fluid (SBF). As fabricated, the 13-93 scaffolds showed a compressive strength of 142 ± 20 MPa, comparable to the values for human cortical bone, while the 13-93B3 scaffolds showed a compressive strength of 65 ± 11 MPa, far higher than the values for trabecular bone. When immersed in SBF, the borate 13-93B3 scaffolds converted faster than the silicate 13-93 scaffolds to a calcium phosphate material, but they also showed a sharper decrease in strength. Based on their high compressive strength, bioactivity, and microstructure favorable for supporting tissue ingrowth, the scaffolds fabricated in this work by robocasting could have potential application in the repair of loaded and non-loaded bone.

INTRODUCTION

There is a great clinical need for the development of low-cost synthetic scaffolds that replicate the porosity, bioactivity, strength, and load-bearing ability of living bone. Bioactive glasses are promising scaffold materials for bone tissue engineering applications because of their unique ability to convert to hydroxyapatite (HA) in vivo, in addition to their proven osteoconductivity and their ability to bond firmly with bone and soft tissues.[1-3]

The silicate bioactive glass designated 45S5 as well as silicate bioactive glasses and glass ceramics based on the 45S5 composition have received considerable interest recently for applications in bone repair.[2] In our previous work, the silicate bioactive glass designated 13-93 (Table I) was used predominantly in the creation of scaffolds because of its ability to be sintered to almost full density without crystallization of the glass.[4] Borate bioactive glasses have been developed recently for potential tissue engineering applications.[5] In particular, the borate glass designated 13-93B3 (Table I), obtained from silicate 13-93 by replacing the molar concentration of SiO_2 by B_2O_3, has been prepared and used in our previous work.[6,7] Because of its lower chemical durability, borate 13-93B3 bioactive glass converts faster and more completely to a hydroxyapatite (HA)-like material than silicate 13-93 glass.[6] The conversion of borate bioactive glass to HA appears to follow a process similar to that for silicate glass, but without the formation of an SiO_2-rich layer.[8] Borate 13-93B3 bioactive glass has been shown to support faster bone formation than silicate 13-93 bioactive glass in a rat calvarial model.[9]

A variety of methods has been used to create porous three-dimensional (3D) scaffolds of polymeric and inorganic biomaterials,[10-12] including rapid prototyping or solid freeform fabrication techniques such as sterelithography[13-14], 3D printing[15], fused deposition modeling[16], extrusion fabrication[17-18] and robocasting.[19-21] Previous work has shown that solid freeform fabrication techniques such as freeze extrusion fabrication and robocasting can be used to create silicate 13-93 and 6P53B bioactive glass scaffolds with compressive strengths comparable to those of human cortical bone.[15,17] However, the creation of borate bioactive glass scaffolds using solid freeform fabrication techniques has received little attention. Ideally, fabrication methods that rely on the use of aqueous solvents are not recommended for the creation of borate bioactive glass scaffolds because of the higher reactivity of borate glass in aqueous media.

In the present work, organic-based suspensions of bioactive glass particles were prepared and used in the creation of 3D scaffolds by robocasting. Scaffolds of silicate 13-93 and borate 13-93B3 glasses with a grid-like microstructure were created and evaluated in vitro to characterize their microstructure, mechanical response in compression, and conversion to HA in a simulated body fluid.

EXPERIMENTAL

Preparation of Bioactive Glass Inks

Silicate 13-93 and borate 13-93B3 bioactive glass frits with the composition given in Table I were supplied by Mo-Sci Corp., Rolla, MO, USA. In the preparation of the slurries (inks) for robocasting, ethyl cellulose (Acros Organics, USA) and polyethylene glycol (PEG 300; Alfa Aesar, Ward Hill, MA) were used as the organic additives (binder/plasticizer), while anhydrous ethanol (Sigma Aldrich, USA) was the solvent.

Table I. Composition of 13-93 and 13-93B3 glasses (in wt%) used in this work; for comparison, the composition of 45S5 bioactive glass is also shown.

Glass	SiO_2	B_2O_3	CaO	Na_2O	K_2O	MgO	P_2O_5
13-93	53		20	6	12	5	4
13-93B3		56.6	18.5	5.5	11.1	4.6	3.7
45S5	45		24.5	24.5			6

Particles of each bioactive glass were prepared by grinding the as-received glass frits for 3 min in a SPEX mill (Model 8500, Metuchen, NJ), sieving to obtain particles of size <100 μm, followed by attrition-milling for 2.5 hours using deionized water (for 13-93) or ethanol (for 13-93B3) as the solvent and zirconia balls (3 mm) as the milling media. The slurries were dried at 60 °C and the powder was sieved through a 53 μm stainless steel sieve to eliminate the agglomerates resulting from the drying step. Particle size analysis showed a median diameter of 2.2 μm for the silicate 13-93 glass and 2.1 μm for borate 13-93B3 glass.

An organic solvent-based system was developed for the preparation of the bioactive glass inks. Concentrated bioactive glass inks with the compositions given in Table II were prepared and used for robocasting. Ethyl cellulose was first dissolved in the solvent, followed by PEG, and the solution was stirred overnight at 25 °C. After adding the bioactive glass particles to the polymer solution, the slurry was mixed for 10 min in a planetary centrifugal mixer (Thinky AR 310).

Table II. Composition of inks (in vol%) used for preparing silicate 13-93 and borate 13-93B3 constructs by robocasting.

Component	Function	13-93	13-93B3
Glass particles	Solid phase	45.0	40.0
Ethyl cellulose	Binder	20.1	12.8
Poly(ethylene glycol) PEG	Plasticizer	6.8	4.3
Ethanol	Solvent	28.1	42.9

Viscosity of the Bioactive Glass Suspensions

The viscosity of the polymer solutions (without bioactive glass particles) and suspensions with silicate 13-93 and borate 13-93B3 particles was measured using a controlled rate rotational viscometer (Haake Viscotester E, Haake Inc., Paramus , NJ) with a small sample adaptor. Measurements were performed with TR series spindles, using 8–13 ml of suspension. Variation in shear stress and the viscosity of the inks were measured as a function of shear rate. All measurements were performed at 25 °C. A specially designed solvent trap was utilized during the measurements to reduce solvent evaporation.

Robocasting

Periodic lattices were created using a robotic deposition (robocasting) apparatus (3D Inks; Stillwater, OK). For the deposition, the ink was housed in a 3 ml syringe and deposited through a tapered stainless steel nozzle (inner diameter, $D = 410$ μm) held in a plastic housing (EFD precision tips, East Providence, RI) at a volumetric flow rate required to maintain a constant x-y table speed of 10 mm/s. Deposition was performed in a closed chamber with an ethanol rich environment to control the drying rate of the extruded material during the printing step.

Binder Removal and Sintering

After printing, the scaffolds were dried under ambient conditions for 24 h, followed by a controlled heat treatment process. Binder burn out was performed in flowing oxygen using very low heating rates in the range 0.1–1°C/min. Sintering was performed for 1 h at 690 °C for silicate 13-93 and at 560 °C for borate 13-93B3 scaffolds, using a heating rate of 5 °C/min. The microstructure of the fabricated scaffolds was examined using scanning electron microscopy; SEM (S-4700, Hitachi, Tokyo, Japan) at an accelerating voltage of 15 kV and a working distance of 12 mm. X-ray diffraction, XRD (Philips XPert) was used to check for the presence of any crystalline phase in the sintered scaffolds. The sintered constructs were ground to a powder (particle size <45 μm), and analyzed using Cu K$_\alpha$ radiation at a scanning rate of 0.01°/min in the 2θ range of 3–90°.

In vitro Bioactivity

The bioactivity of the sintered scaffolds was evaluated in vitro in simulated body fluid (SBF) described elsewhere.[22] The conversion of the bioactive glass to HA is accompanied by a weight loss, and this weight loss was measured to determine the rate and the extent of the conversion. A ratio of 1 g of scaffold to 100 mL of SBF was used in the conversion experiments. Scaffolds of each glass were each immersed in a polyethylene bottle containing the SBF solution, and kept for up to 30 days without shaking in an incubator at 37 °C. Four scaffolds were used for each immersion time. After removal from the SBF, the scaffolds were dried at 60 °C and weighed. SEM and XRD were used to analyze the structure of the reacted scaffolds, using the conditions described previously.

Mechanical Testing

The compressive strengths of cube-shaped scaffolds (length = 7.5 mm), as fabricated, and after immersion in SBF, were measured using a mechanical testing machine (Model 4881; Instron, Norwood, MA) at a deformation rate of 0.5 mm/min. Five samples in each group were tested, and the data were expressed as a mean ± standard deviation. Scaffolds immersed in SBF were tested as removed from the SBF without drying.

RESULTS AND DISCUSSION

Rheology of Bioactive Glass Inks

Figure 1 shows the viscosity of silicate 13-93 and borate 13-93B3 glass inks as a function of shear rate. Both suspensions showed a desirable shear thinning behavior and the viscosity data can be well fitted by the Herschel-Bulkley model.[20] The viscosity of the polymer solutions (prior to the addition of the bioactive glass particles) had lower viscosity values (at any shear rate) and showed Bingham plastic flow behavior (results not shown). Addition of glass particles to the polymer solution resulted in an increase in viscosity and transition to the shear thinning behavior shown in Fig. 1.

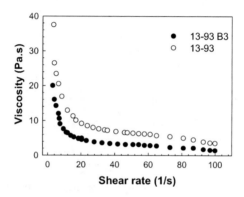

Figure 1. Viscosity of silicate 13-93 and borate 13-93B3 bioactive glass inks used in robotic deposition as a function of shear rate.

Bioactive Glass Scaffolds

Figure 2 shows optical images of silicate 13-93 and borate 13-93B3 scaffolds with a cubic geometry prepared in this work using the robocasting technique. The external shape and grid-like microstructure formed in the robocasting step (Figs. 2a, 2d) were retained after the binder burnout and sintering steps (Figs. 2b, 2e). After sintering, the glass struts (filaments) were almost fully dense (Fig. 2c, 2f). The linear shrinkage of the scaffolds during sintering was almost isotropic, in the range 25–30%. The sintered scaffolds had a pore width and strut diameter of ~400 μm and 290 ± 10 μm, respectively, and a porosity of 48 ± 3%, as determined from the mass and external dimensions. The interconnectivity of the pores in the scaffolds was maintained after sintering, and the pore width was in the range shown to be favorable for tissue ingrowth in vivo.

In vitro Bioactivity

Bioactive glasses upon implantation or in vitro conditions convert to an amorphous calcium phosphate or hydroxyapatite (HA)-like material, which is responsible for their strong bonding with surrounding tissue.[2,3,6,7] In the present study, the bioactivity of the sintered silicate and borate scaffolds were tested by immersion in SBF. After immersion of the scaffolds for 30 days, the weight loss of the borate 13-93B3 scaffolds was 42 ± 7%. Assuming the glass is converted completely to HA, the theoretical weight loss is calculated to be 67%. The lower value of the measured weight compared to the theoretical value indicates that the borate 13-93B3 scaffolds are not completely converted to HA within the 30 day immersion. Experiments are in progress for longer immersion times. Under the same conditions (30 days in SBF), the weight loss of the silicate 13-93 scaffolds was only $5 \pm 2\%$ (Table III).

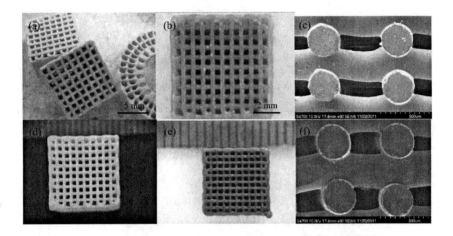

Figure 2.(a) Image of 13-93 scaffold after robocasting; (b) Image of 13-93 after sintering; (c) SEM of cross section of sintered 13-93 scaffold showing dense struts; (d) Image of 13-93B3 scaffold after robocasting; (e) Image of 13-93B3 after sintering; (f) SEM of cross section of sintered 13-93B3 scaffold showing dense struts.

The pH of the SBF (starting value = 7.4) increased with time upon immersion of the bioactive glass scaffolds. The increase in the pH with immersion time showed trends that corresponded to those of the weight loss data; the pH at any given immersion time was higher for the borate 13-93B3 scaffolds. After immersion of the scaffolds for 30 days, the pH of the SBF increased to 8.3 for the silicate 13-93 scaffolds and to 8.7 for the borate 13-93B3 scaffolds (Table III).

The results showed that the conversion of the borate 13-93B3 scaffolds was much faster than the silicate 13-93 scaffolds, which is consistent with the results of previous studies.[6,8,23] The far higher conversion rate of some borate bioactive glasses (such as 13-93B3), when compared to silicate bioactive glasses, has been discussed in terms of the lower chemical durability of the borate glasses. Based on the weight loss data, approximately two-thirds of the borate 13-93B3 scaffolds was converted after immersion for 30 days in SBF. Previous work has shown complete conversion of particles (150–300 μm) and highly porous constructs (prepared by a foam replication technique) within 7 days.[6,24] The conversion rate of a given bioactive glass in an aqueous phosphate solution depends on several factors, such as the surface area and geometry of the glass, as well as the phosphate ion concentration and pH of the solution. In the present work, the coarse dense struts (400 μm), coupled with the low phosphate ion concentration of the SBF, could be key factors responsible for the incomplete conversion of the 13-93B3 scaffolds within a thirty-day period.

Table III. Characteristics of as-fabricated bioactive glass scaffolds and after immersion for 30 days in a simulated body fluid.

Scaffold	Porosity (%)	Strut diameter (μm)	Pore width (μm)	Weight loss	Final pH of SBF*
13-93	48 ± 3	400	290 ± 10	5 ± 2	8.3
13-93B3	48 ± 3	400	290 ± 10	42 ± 7	8.7

*Starting pH = 7.4.

SEM images in Fig. 3 show the surfaces of the silicate 13-93 and borate 13-93B3 scaffolds after immersion in SBF for 30 days. Compared to the smooth surface of the sintered scaffolds (see Fig. 2), the surface of the reacted scaffolds had a fine particulate structure. The morphology of the particles on the reacted surfaces was needle-like for the 13-93 scaffolds and more rounded for the 13-93B3 scaffolds. These morphologies are typical of HA-like material formed by the conversion of silicate and borate bioactive glasses in an aqueous phosphate solution.[6,25]

Figure 3. SEM images of the surface of silicate 13-93 scaffolds (a, b) and borate 13-93B3 scaffolds (c, d) after immersion for 30 days in SBF; (a, c) lower magnification, (b, d) higher magnification.

XRD analysis of the silicate 13-93 and borate 13-93B3 scaffolds after reaction in SBF for 30 days did not show the presence of a crystalline phase. Fourier transform infrared (FTIR) spectroscopy showed resonances corresponding to a calcium phosphate (results not shown). Taken together, the XRD and FTIR analyses indicate that the converted layer might be an amorphous calcium phosphate layer, a precursor to the formation of the crystalline HA phase. For the borate 13-93B3 scaffolds reacted in SBF for 30 days, heat treatment (1 h at 900 °C) resulted in the formation of peaks in the XRD pattern which corresponded a reference HA (JCPDS 72-1243) (Fig. 4). The HA peaks presumably resulted from the crystallization of the amorphous calcium phosphate formed during the conversion reaction. The XRD pattern also showed minor peaks corresponding to the presence of calcium metaborate, $Ca(BO_2)_2$ (JCPDS 78-1277) presumably due to crystallization of the unconverted glass in the scaffold.

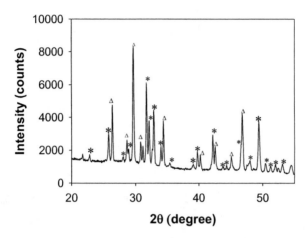

Figure 4. XRD patterns of borate 13-93B3 scaffolds after immersion in SBF for 30 days followed by heating for 1 h at 900°C. (*Hydroxyapatite, JCPDS 72-1243; Δ Calcium metaborate, JCPDS 78-1277).

Mechanical Response

Both the silicate 13-93 and the borate 13-93B3 scaffolds showed an elastic mechanical response in compression, followed by failure in a brittle manner. The compressive strength of the as-fabricated silicate 13-93 scaffolds (after sintering) was 142 ± 20 MPa, which is in the range 100–150 MPa reported for human cortical bone.[26] In comparison, the compressive strength of the as-fabricated borate 13-93B3 scaffolds was 65 ± 11 MPa. The lower strength of the borate scaffolds is consistent with previous work[6] in which the compressive strength of silicate 13-93 and borate 13-93B3 scaffolds with a trabecular microstructure was measured. For the same microstructure, the difference in strength between the silicate and borate scaffolds may be related to the strength of the Si–O and B–O bonds in the individual glasses.

Upon immersion in SBF, the compressive strength of both groups of scaffolds decreased with immersion time (Fig. 5). The borate 13-93B3 scaffolds showed a faster decrease in strength when compared to the silicate 13-93 scaffolds, a trend that is consistent with the results of previous work for 13-93 and 13-93B3 scaffolds with a trabecular microstructure.[6] After immersion in SBF for 30 days, the mean compressive strength of the silicate 13-93 scaffolds decreased by ~40% to 87 ± 28 MPa. In comparison, the mean compressive strength of the borate 13-93B3 scaffolds decreased by ~85% to 9 ± 4 MPa.

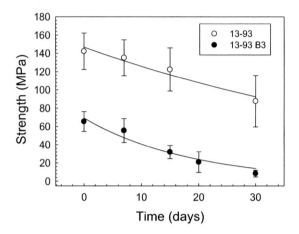

Figure 5. Compressive strength of silicate 13-93 and borate 13-93B3 scaffolds as a function of immersion time in SBF.

CONCLUSIONS

Organic based suspensions (inks) were developed and utilized for the robotic deposition of porous three-dimensional scaffolds of silicate 13-93 and borate 13-93B3 glass with a grid-like microstructure. The as-fabricated scaffolds (porosity ≈50%) consisted of dense glass filaments of diameter ≈400 μm) and interconnected pores of width ≈300 μm. As fabricated, the silicate 13-93 scaffolds had a compressive strength of 142 ± 20 MPa, which decreased to 87 ± 28 MPa after immersion for 30 days in a simulated body fluid (SBF). In comparison, the as-fabricated borate 13-93B3 had a compressive strength of 65 ± 11 MPa, which decreased to 9 ± 4 MPa after immersion for 30 days in SBF. After 30 days in SBF, the weight loss of the silicate 13-93 and borate 13-93 scaffolds was 5 ± 2% and 42 ± 7%, respectively, indicating a far faster conversion rate of the borate glass scaffolds. The conversion product consisted of an amorphous calcium phosphate material after the thirty-day immersion. Based on their high compressive strength, bioactivity, and microstructure favorable for supporting tissue ingrowth, the scaffolds fabricated in this work by robocasting could have potential application in the repair of loaded and non-loaded bone.

Acknowledgements

The authors would like to thank Hailuo Fu and Xin Liu for technical assistance, and Mo-Sci Corp., Rolla, MO, for kindly supplying the bioactive glasses used in this work. Support for this research was provided by the Center for Bone and Tissue Repair and Regeneration at Missouri S&T, and by The Scientific and Technical Research Council of Turkey in the form of a TUBITAK-BIDEB 2219 fellowship to A. Deliormanlı.

REFERENCES

1. L-C Gerhardt, A.R. Boccaccini., Bioactive Glass and Glass-Ceramic Scaffolds for Bone Tissue Engineering, *Materials*, **3**, 3867-3910 (2010).
2. L.L. Hench, Bioceramics, *J Am Ceram Soc*, **81**, 1705-1728 (1998).
3. M.Brink, T.Turunen, R.Happonen, A. Yli-Urppo, Compositional Dependence of Bioactivity of Glasses in the System $Na_2O-K_2O-MgO-CaO-B_2O_3-P_2O_5-SiO_2$, *J Mater Sci Mater Med*, **37**,114–21 (1997).
4. Q.Fu, M. N. Rahaman, B. S. Bal, K. Kuroki, R. F. Brown, In Vivo Evaluation of 13-93 Bioactive Glass Scaffolds with Trabecular and Oriented Microstructures in a Subcutaneous Rat Implantation Model. *J Biomed Mater Res Part A*, **95A**, 235-244 (2010).
5. D.E.Day, J.E.White, R.F.Brown, K.D.McMenamin, Transformation of Borate Glasses into Biologically Useful Materials, *Glass Technol Part A*, **44**,75–81 (2003).
6. Q.Fu., M.N.Rahaman, H.Fu, X.Liu, Silicate, Borosilicate, and Borate Bioactive Glass Scaffolds with Controllable degradation rate for bone tissue engineering applications. I. Preparation and in vitro degradation, *J Biomed Mater Res Part A*, **95A**, 164–171 (2010).
7. Q.Fu, M. N. Rahaman, B. S. Bal, L. F. Bonewald, K. Kuroki, R.F. Brown, Silicate, Borosilicate, and Borate Bioactive Glass Scaffolds with Controllable Degradation Rate for Bone Tissue Engineering Applications. II. In vitro and in vivo biological evaluation, *J Biomed Mater Res Part A,* **95A,** 172–179 (2010).
8. W.Huang , D.E.Day, K.Kittiratanapiboon, M.N.Rahaman, Kinetics and Mechanisms of the conversion of silicate (45S5), Borate, and Borosilicate glasses to hydroxyapatite in dilute phosphate solution, *J Mater Sci Mater Med*, **17**, 583–96 (2006).
9. M.N.Rahaman, D.E.Day, B.S.Bal, Q.Fu, S.B.Jung, L.F.Bonewald, Bioactive Glass in Tissue Engineering, *Acta Biomater*, **7**, 2355-2373 (2011).
10. H.Fu, Q.Fu, N.Zhou, W.Huang, M.N.Rahaman, D.Wang, X.Liu, In Vitro Evaluation of Borate-based Bioactive Glass Scaffolds Prepared by a Polymer Foam Replication Method, *Mater Sci Eng C*, **29**, 2275–2281 (2009).
11. B.Chen, Z.Zhang, J.Zhang, M.Dong, D.Jiang, Aqueous Gel-Casting of Hydroxyapatite, *Mater Sci Eng A*, **435-436**, 198-203 (2006).
12. X.Liu, M.N.Rahaman, Q. Fu, Oriented Bioactive Glass (13-93) Scaffolds with Controllable Pore Size by Unidirectional Freezing of Camphene based Suspensions: Microstructure and Mechanical Response, *Acta Biomater*, **7**, 406-416 (2011).
13. W-Y.Yeong, C-K.Chua, K-F.Leong, M.Chandrasekaran, Rapid Prototyping in Tissue Engineering: Challenges and Potential, *Trends Biotechnol*, **22**, 643-652 (2004).
14. B.Stevens, Y.Yang, A.Mohandas, B.Stucker, K.T.Nguyen, A Review of Materials, Fabrication Methods, and Strategies Used to Enhance Bone Regeneration in Engineered Bone Tissues, *J Biomed Mater Res Part B*, **85B**, 573-582 (2007).
15. D.L.Cohen, E.Malone, H. Lipson, L.Bonassar, 3D Direct Printing of Heterogeneous Tissue Implants, *Tissue Eng.*, **12**, 5, 1325-1335 (2006).
16. I. Zein, D.W. Hutmacher, K. C.Tan, S.H.Teoh, Fused Deposition Modeling of Novel Scaffold Architectures for Tissue Engineering Applications, *Biomaterials*, **23**, 1169–1185 (2002).
17. N.D.Doiphode, T.Huang, M.C.Leu, M.N.Rahaman, D.E.Day, Freeze Extrusion Fabrication of 13-93 Bioactive Glass Scaffolds for Bone Repair, *J Mater Sci Mater Med*, **22**, 515-523 (2011).
18. X. Lu, Y.Lee, S.Yang, Y.Hao, J.R.G. Evans, C.G.Parini, Solvent Based Paste Extrusion Solid Freeforming, *J Euro Ceram Soc*, **30**, 1-10 (2010).
19. Q.Fu, E.Saiz, A.Tomsia, Bio-inspired Highly Porous and Strong Glass Scaffolds, *Adv Funct Mater*, **21**, 1058-1063 (2011).

20. J.Russias, E.Saiz, S.Deville, K.Gryn, G.Liu, R.K. Nalla, Tomsia A.P., Fabrication and In Vitro Characterization of Three Dimensional Organic/Inorganic Scaffolds by Robocasting, *J Biomed Mater Res Part A*, **83A**, 434-445 (2007).
21. S. Michna, W. Wu, J.A. Lewis, Concentrated Hydroxyapatite Inks for Direct-Write Assembly of 3-D Periodic Scaffolds, *Biomaterials* **26**, 5632-39 (2005).
22. T. Kokubo, H.Kushitani, S.Sakka, T.Kitsugi, T.Yamamuro, Solutions able to Reproduce In Vivo Surface-Structure Changes in Bioactive Glass-Ceramic A-W. *J Biomed Mater Res*, **24**,721–734 (1990).
23. S.B. Jung, D. E. Day, Conversion Kinetics of Silicate, Borosilicate, and Borate Bioactive Glasses to Hydroxyapatite, *Phys Chem Glasses: Eur J Glass Sci Technol B*, **50**, 85–88 (2009).
24. A. Yao, D.Wang, W.Huang, Q.Fu, M.N.Rahaman, D.E.Day, In Vitro Bioactive Characteristics of Borate-based Glasses with Controllable Degradation Behavior, *J Am Ceram Soc*, **90**, 303–306 (2007).
25. Q.Fu, M.N. Rahaman, B. S.Bal, R. F. Brown, D. E. Day, Mechanical and in vitro Performance of 13–93 Bioactive Glass Scaffolds Prepared by a Polymer Foam Replication Technique, *Acta Biomater,* **4**, 1854–1864 (2008).
26. J.Y.Rho, Mechanical Properties and the Hierarchical Structure of Bone, *Med Eng Phys*, **20**, 92-102 (1998).

TRANSLUCENT ZIRCONIA-SILICA GLASS CERAMICS FOR DENTAL CROWNS

Wei Xia, Cecilia Persson, Erik Unosson, Ingrid Ajaxon, Johanna Engstrand, Torbjörn Mellgren, Håkan Engqvist
Division for Applied Materials Science, Department of Engineering Sciences, Uppsala University
Uppsala, Sweden

ABSTRACT

Lithium disilicate glass ceramics are commonly used in dental crowns because of their optical and mechanical properties. Alternative materials, such as zirconia-silica glass ceramics, with improved mechanical properties may be of interest for more demanding applications. In this study, a sol-gel method was optimized to produce nano grain-sized glass ceramics based on the zirconia-silica system. The zirconia-silica glass ceramic was translucent and a higher fracture toughness could be achieved in comparison to lithium disilicate (IPS e.max CAD) via a proper heat treatment. In conclusion, a translucent glass ceramic system with promising mechanical properties was prepared by a sol-gel method, and has a good potential for dental applications.

INTRODUCTION

Ceramics and glass ceramics are commonly used as dental materials such as crowns due to their adequate mechanical properties and appealing aesthetics. Because of their good mechanical properties, all-ceramic materials are commonly used for dental applications. However, their natural opacity makes them difficult to adapt to the colour of the surrounding teeth. Glass ceramics, such as lithium disilicates, can be translucent and more aesthetically pleasing. However, the glass ceramics generally have a fracture toughness and flexural strength less than half that of the all-ceramics [1]. A translucent glass-ceramic with improved mechanical properties would therefore be interesting for dental restoration applications.

Zirconia ceramics are known for their high fracture toughness and good mechanical strength. However, they are generally not translucent, and do not fulfill the aesthetical requirements. A potential alternative may be a glass-ceramic of zirconia-silica. The melting method that is commonly used to prepare lithium silicate glass ceramics is not good for preparation of zirconia-silica system because crystallized zirconia makes the sample opaque. A low-cost sol-gel technique has been reported by Nogami et al. [2] to produce translucent, high toughness glass ceramics in the ZrO_2-SiO_2 system. The fracture toughness could reach up to 4.8 MPa√m, in comparison to the 4.2 MPa√m reported for the lithium disilicate IPS E.max Press [3]. The high fracture toughness was attributed to the transformation toughening mechanism of tetragonal - monoclinic zirconia.

Therefore, the ZrO_2-SiO_2 glass ceramics could be evaluated for dental applications. However, the fracture toughness has only been evaluated for 60% ZrO_2 [2]. No data on the elastic modulus is available, which is important for dental applications [4]. Furthermore, a glass ceramic with nano-grain size could have a higher mechanical strength because of the toughening effect of the residual stress in the grain boundary.

In this study, a sol-gel method was optimized and used to prepare a translucent, nano-grain size glass ceramic suitable for dental crowns.

EXPERIMENTAL

Materials

Alkoxide precursors tetraethyl orthosilicate (TEOS) and 70 wt.% tetrapropyl zirconate (TPZ) in 1-propanol (all chemicals were acquired from Sigma-Aldrich, St Louis, MO, USA) were used to prepare ZrO_2 - SiO_2 glass ceramics of 30mol% ZrO_2, using a sol-gel method adapted from Nogami et al. [2]. A drying control additive was used to reduce the number of cracks in the specimens.

The preparation was initiated by mixing ethanol (EtOH, >95 %), aqueous hydrochloric acid (HCl) and the drying control additive dimethylformamide (DMF) in a 50 ml round bottom flask, followed by addition of TEOS under continuous stirring. The resulting molar ratio of the solution was 1:1:1:1 - TEOS:DMF:EtOH:H_2O. The partially hydrolysed TEOS was then magnetically stirred for 3 h in order to obtain a clear sol. The desired amount of TPZ was then added slowly using a micropipette and magnetic stirring of the solution was continued overnight. As EtOH is volatile, the sol was always kept covered during mixing in order to minimize any changes in concentration due to evaporation. 0.15 M HCl was then added drop by drop to initiate the final hydrolysis and polymerization of a monolithic gel. The sol was transferred to Teflon moulds of 25 mm diameter, which were sealed with polymer film for controlled evaporation. The sols were left to gel and age until approximately 40% shrinkage was observed and a stiff gel was formed. After formation of a stiff xerogel, samples were moved to an oven and kept at 100°C in an atmosphere of 100% relative humidity for 5 h. The temperature was then raised to 150 °C and held for 15 h. The subsequent heat treatment process was initiated with a calcination plateau at 800 °C with a holding time of one hour. Sintering of the samples was then carried out at 1100 °C, with holding times of 10 or 15 h. The ramp rate was 20-30 °C/h.

CHARACTERIZATION

Thermal analysis

The as-prepared xerogel was analyzed with a simultaneous DTA/TGA (Differential Thermal Analysis and Thermal Gravimetric Analysis) machine (NETZSCH, STA 409 CD). DTA/TGA was run from room temperature to 1100 °C with a ramp rate of 0.5 °C/min in air.

Crystallinity

The crystallinity of the samples was analyzed by X-ray diffraction analysis using a Siemens Diffractometer D5000 (Siemens AG, Munich, Germany). The analysis was performed on bulk samples over the 2θ range of 10 to 80° and at a step size of 0.05 and scanning speed of 5 s/step. The particle size was calculated using Scherrer's formula.

Mechanical properties

Preparation of samples for nano- and microindentation included mounting in thermoplastic resin, wet grinding and polishing. Grit papers from 320 to 1200 were used and polishing was done using diamond pastes from 6 μm through to 1 μm.

Young's modulus and hardness on the nano scale were assessed with an Ultra Nanoindenter (CSM Instruments SA, Peseux, Switzerland). A load of 8 mN at a speed of 8mN/min was applied on the sample surface. For each sample, 10-20 indentations were done with a distance of 40μm between adjacent indentations. Poisson's ratio was set to 0.3 as literature data for Poisson's ratio for the ZrO_2-SiO_2 system and similar ceramics ranges between 0.25-0.3 [1, 5, 6] and the difference in Poisson's ratio would affect the elastic modulus to an extent less than 0.3%. The Oliver-Pharr method [7] was used to calculate the Young's modulus.

Indentation on the micro scale was done using a micro hardness tester (Buehler Micromet 2104, Lake Bluff, IL, USA) equipped with a Vickers diamond. The DIN EN 843-4 standard [8] was used for the test. A load of 2 kg was used for making at least 10 indentations per sample. Indentation diagonals were measured optically using an Olympus AX70 light microscope (Olympus Corp., Tokyo, Japan), at a magnification of 50 times. Tip cracks were measured by generating a picture from a CCD camera connected to the microscope (Lumenera Corp., Ottawa, ON, Canada), together with the associated software. Cracks were measured within approximately 30 minutes from when the applied load was removed.

Fracture toughness was calculated from the Young's modulus obtained from the nano indentations and the Vickers hardness obtained from the micro indentation, together with average crack lengths as measured optically. The equation proposed by Niihara et al. [9] for Palmqvist cracks ($l/a<2.5$) was used for the calculations:

$$K_c = 0.018H_v \sqrt{a}\left(\frac{E}{H_v}\right)^{2/5}\left(\frac{l}{a}\right)^{-1/2}$$ (4)

where
a is the indentation diagonal
E is the Young's modulus,
H_v is the Vickers hardness
l the crack length from the indent tip.

The crack profile was characterized by step-by-step polishing away the Vickers indent using a 1 μm diamond paste and observing the indentation profile under the microscope. Since the cracks could be polished away while the indentation mark was still visible, *i.e.* the cracks did not extend under the indentation, the cracks were established to be in the Palmqvist crack system. Measurements were made on our samples, as well as a reference specimen of IPS e.max CAD.

Statistical analysis

Statistical analysis was performed using IBM SPSS Statistics 19.0. Since homogeneity of variances could not be ensured for all variables (Levene's test), Welch's test of equality of means was used, together with Tamhane's multiple comparison test at a significance level of 0.05.

RESULTS AND DISCUSSION
Thermal analysis

In the DTA/TGA diagram (Figure 1), for 30 % zirconia content, no glass transition temperature was apparent, but a crystallization of the zirconia could be seen around 900 °C which is comparable with results from Aguilar et al [10]. According to McPherson [11] the glass transition temperature of both pure SiO_2 and pure ZrO_2 is around 1200 °C, which suggests that the glass transition temperature may not be apparent in our DTA. The slight weight gain from 650 to 1100 °C could be due to the reaction of the materials with oxygen in air.

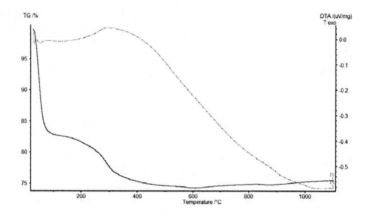

Figure 1 DTA/TGA diagram for the zirconia-silica xerogel.

Crystallinity and translucency

Optical observation showed that the samples after sintering could still appear translucent (Figure 2). Bulk XRD analysis was performed on samples sintered at 1100 °C for 10 or 15 h. Only tetragonal ZrO_2 peaks were found, along with SiO_2 in some instances, as shown in Figure 3. It is in accordance with previous studies where a sintering temperature of 1100 °C has been used [2, 12]. The peaks of SiO_2 point to tridymite. According to the calculation from Scherrer's formula, the crystallite size for samples sintered for 10 and 15 hours was approximate 29 and 33 nm, respectively. The monoclinic phase could also be observed if the temperature was higher than 1150 °C (data not shown).

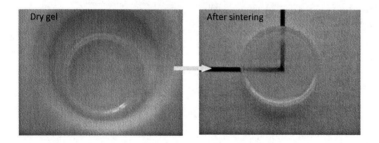

Figure 2 Optical images of the Zirconia-silica system: dry gel and sintered sample.

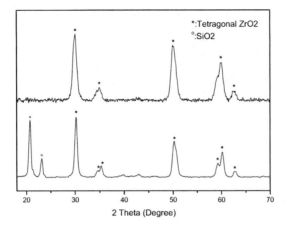

Figure 3 XRD analyses of samples with a sintering time of 10 and 15 hours.

Mechanical properties

The Young's modulus (E), nano- and microhardness (H) and the fracture toughness (Kc) resulting from the nano- and microindentations are shown in Figure 4. The samples with a sintering time of 15 h showed a high fracture toughness, an average of 4.3 MPa√m, compared to the average 3.1 MPa√m of the IPS e.max CAD. This is somewhat lower than that of Nogami and Tomozawa [2], who produced samples containing twice the amount of zirconia. However, the production of samples containing such high amounts of zirconia proved difficult, due to increased risk for precipitation during preparation. Future work will investigate the possibilities of producing such material. The statistical test of equality of means found significant differences for all output parameters ($p \leqslant 0.001$). Tamhane's multiple comparison test revealed a statistically significant difference between all groups (e.max, zirconia-silica sintered at 10 h and zirconia-silica sintered for 15 h) for the elastic modulus ($p < 0.001$). IPS e.max was found to have a significantly lower hardness (both nano and micro) than the zirconia-silica glass ceramics ($p < 0.001$), while the two zirconia-silica formulations were not statistically different in terms of hardness. For the fracture toughness, the zirconia-silica glass ceramic sintered for 15 h gave significantly higher Kc than IPS e.max. No statistically significant difference was found between zirconia-silica sintered at 10 h and IPS e.max, nor between the two zirconia-silica formulations. Presently, there are two main possible toughening mechanisms: (1) tetragonal-monoclinic transformation, (2) residual stresses in the ceramic grain boundary in a ZrO_2-SiO_2 system with nano-grain size [13]. The nano-sized ZrO_2 tetragonal phase could be stabilized by the silica matrix due to its constraint and the formation of a Zr-O-Si interlayer surrounding the ZrO_2 grains. Future work will further evaluate these two possibilities.

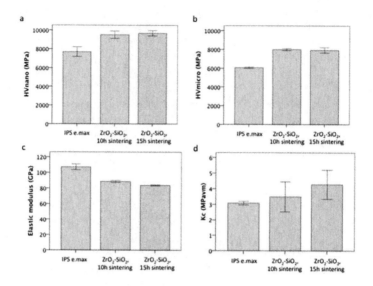

Figure 4 Nano- (a) and micro-hardness (HV) (b), Young's modulus (E) (c) and the fracture toughness (Kc) (d) obtained from the nano- and micro-indentations.

CONCLUSIONS

Translucent zirconia-silica glass ceramics with high strength have been produced using a sol-gel method. Compared to commercial lithium disilicate glass ceramics, a higher fracture toughness could be achieved. For further development, the content of ZrO_2 in this system could be further optimized. Although promising results were obtained it should be noted that the process is sensitive to outer disturbance because of the formation of cracks during drying.

ACKNOWLEDGEMENT

Financial support from the Swedish research council (VR) and Sweden's Innovation Agency (VINNOVA).

REFERENCES

1.   H. Yilmaz, C. Aydin and B.E. Gul, "Flexural strength and fracture toughness of dental core ceramics." J. Prosthet. Dent., 98(2) 120-8 (2007).
2.   M. Nogami and M. Tomozawa, "ZrO_2-Transformation-Toughened Glass-Ceramics Prepared by the Sol-Gel Process from Metal Alkoxides". J. Am. Ceram. Soc., 69(2) 99-102 (1986).
3.   U. Lohbauer, F.A. Muller and A. Petschelt, "Influence of surface roughness on mechanical strength of resin composite versus glass ceramic materials" Dent. Mater., 24(2) 250-6 (2008).
4.   ISO standard, *ISO 6872: Dentistry - Ceramic materials*, 2008, International Organization for Standardization.

5. A. Makishima and J.D. Mackenzie, "Calculation of bulk modulus, shear modulus and Poisson's ratio of glass" J. Non-Cryst. Solids, 17(2) 147-57 (1975).
6. M. Guazzato, K. Proos, L. Quach and M.V. Swain, "Strength, reliability and mode of fracture of bilayered porcelain/zirconia (Y-TZP) dental ceramics" Biomaterials, 25(20) 5045-52 (2004).
7. W.C. Oliver and G.M. Pharr, "An improved technique for determining hardness and elastic modulus using load and displacement sensing indentation experiments" J. Mater. Res., 7(6) 1564-83 (1992).
8. DIN, *DIN EN 843-4: Mechanical properties of monolithic ceramics at room temperature*, 2005.
9. K. Niihara, R. Morena and D.P.H. Hasselman, "Evaluation of K_{Ic} of brittle solids by the indentation method with low crack-to-indent ratios" J. Mater. Sci. Lett., 1(1) 13-6 (1982).
10. D.H. Aguilar, L.C. Torres-Gonzalez, L.M. Torres-Martinez, T. Lopez and P. Quintana, "A study of the crystallization of ZrO_2 in the sol-gel system: ZrO_2-SiO_2" J. Solid State Chem., 158(2) 349-57 (2001).
11. R. McPherson, "Formation of tetragonal ZrO2-glass structures in rapidly quenched Al_2O_3-SiO_2-ZrO_2" J. Mater. Sci. Lett., 6(7) 795-6 (1987).
12. I.M. Miranda Salvado, C.J. Serna and J.M. Fernandez Navarro, "ZrO_2-SiO_2 materials prepared by sol-gel" J. Non-Cryst. Solids, 100(1-3) 330-8 (1988).
13. S.W. Wang, X.X. Huang and J.K. Guo, "Mechanical properties and microstructure of ZrO_2–SiO_2 composite" J. Mater. Sci., 32(1) 197-201. (1997)

USING MICROFOCUS X-RAY COMPUTED TOMOGRAPHY TO EVALUATE FLAWS IN CERAMIC DENTAL CROWNS

Y. Zhang , J. C. Hanan*
School of Mechanical and Aerospace Engineering
Oklahoma State University, Stillwater, OK 74078

ABSTRACT
        The size and distribution of flaws in ceramic dental crowns are important factors that will affect fatigue and fracture. Microfocus X-ray computed tomography (μCT) was used to characterize flaws in three crowns made at different cooling rates to help determine the best manufacturing process. Results show that there are flaws as big as 220 μm in porcelain, and the faster cooling rate corresponds to more flaws but not necessarily big flaws. The crown with the slowest cooling rate had fewer flaws. The corresponding critical fracture stress was predicted based on the biggest flaw from microtomography.

INTRODUCTION
        With improved toughness due to the introduction of materials like zirconia and zirconia composites, all ceramic crowns have become more practical even as they are popular due to their natural color and biocompatibility [1-2]. Ceramics are often very brittle and contain a large number of preexisting flaws which formed during processing due to incomplete densification, anisotropic thermal expansion and modulus between grains for example. These flaws normally distribute unevenly. One ceramic material often displays a range of fracture properties due to variation of flaw size and distribution. In some cases, crowns fail unexpectedly due to flaws. Flaw size and distribution are also the reason for differences between experimentally observed and theoretically predicted critical loads for the initiation of contact-induced radial cracks in brittle coatings on compliant substrates[3-7]. Defects and stress formed during manufacture will accelerate failure. Research has shown there are large residual stresses [8-9]. Lohbauer et al. [4] analyzed fractography of a failed zirconia frame work and found that the defect cluster in the veneer layer and preexisting stress are reasons of failure, and defects at 35 μm can induce local veneer chipping. Reliability and longevity of all ceramic crowns have always been a concern.
        Knowing the flaw size and distribution will be a significant advantage and can help control the quality of crowns or other dental restorations and reduce the failure rate in clinical practice. μCT can be a powerful tool to evaluate the flaws, since it is nondestructive and can successfully visualize and measure internal structure with almost no material preparation. μCT is efficient and convenient for 3D evaluation. This gives it an advantage over conventional optical microscopes and Secondary Electron Microscopes (SEM). The latter needs samples physically cut into different two dimensional cross-sections, and some 3D information is lost during the process. The disadvantage of μCT is its working spatial resolution is lower than SEM. μCT operates the same way as a conventional medical scanner, but with much higher spatial resolution. The μCT technique is based on the principle that X-rays attenuate as they pass through the object. Since Hounsfield [10] first introduced microtomography to medical science, its application has been extended to geosciences, such as, paleontology, soil science, petroleum engineering, and sedimentology [11]. Development of new generations of high resolution μCTs in the last decade [12] makes it possible to characterize mineral distribution [13-14] and pore structures [15-16].
        This study evaluated the flaw size and distribution in three dental crowns made at different cooling rates using μCT and provided information to help determine the optimal manufacturing

* Corresponding author, e-mail: Jay.Hanan@gmail.com

procedure. Faster cooling rates speed production time in an industry where throughput is challenging. Reducing costs and increasing performance both significantly benefit patient needs.

EXPERIMENT

Material Sample Preparation

Three ceramic crowns with a 3% yttiria stabized tetragonal zirconia polycrystalline (3Y-TZP) core (3M LAVA, St. Paul, MN) and CZR porcelain veneer (Noritake CZR, Japan) made at different cooling rates were used to image flaws. 3Y-TZPis a preferable dental core material due to its high fracture toughness and flexural strength. Zirconia can exist in three crystallographic forms based on sintering temperature at ambient pressure. The phases are monoclinic at room temperature to 1170 C, tetragonal at 1170 to 2370 C, and cubic at 2370 and up to the melting point. The tetragonal to monoclinic phase transformation occurs at 950°C during cooling, if no stabilizing agent is present [3]. A second phase of 3% Yttiria is normally used to stabilize the tetragonal zirconia phase. This was also done here. Porcelain was veneered to the zirconia following the typical procedure used by dental labs for zirconia-porcelain crowns. All the manufacturing procedures were the same for the three crowns except the hold time at high temperature and furnace cool time in the final step. The detailed schedule for the final firing step is listed in Table 1. One crown was processed following the normal industry standard cooling procedure, while the other two crowns were made at slower and faster than the normal cooling rate. Figure 1 shows an image of the crown.

Figure 1 Photo of a typical crown sample.

Table 1. List of high temperature hold time and cooling time in the final firing step.

final firing procedure	high temperature (°C)	hold time (min)	furnace cool time (min)
slow	930	2	8
normal	930	0.5	4
fast	930	0.5	0

Microtomography

A high resolution desktop Skyscan 1172 (Skyscan, Kontich, Belgium) μCT equipped with a 10x megapixel camera and a 100kV tungsten source was used to scan the crowns. The nominal resolution was less than 1 μm and spatial resolution was 5 μm. The technique is based on Beer's law that the intensity of X-rays will attenuate as they travel through objects.

$$\frac{I}{I_0} = \exp(-\mu t) \tag{1}$$

Where I_0 is intensity of the incoming X-ray beam, I is the intensity of the attenuated X-ray, μ is the linear attenuation coefficient, and t is the material thickness.

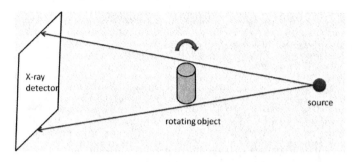

Figure 2 Schematic X-ray μCT setup

The X-ray source was a cone beam. As the crown sample rotates on the high precision stage and X-rays pass through, a series of shadow transmission images were obtained at different angular views. The complete 3D image, representing the internal structure, consists of a stack of 2D cross section images constructed from shadow images using NRecon software (Skyscan, Belgium). A schematic setup of the scanner is shown in Figure 2.

The parameters used for the scan were 100kV power with an exposure time of 474 ms for each frame at a resolution of 14.7 μm. An Al+Cu filter was used to minimize beam hardening. The crowns were scanned at every 1° rotation with four frames for better quality images. It took only 11 minutes to finish scanning one crown. 205 radiographs were obtained and were reconstructed into 484 two-dimensional cross section slices. All three crowns were scanned. Three-dimensional rendering software (Amira) was used to visualize the tomographs. Also a small portion of porcelain on top of the cusp, 2.85mm thick, of the slow cooled crown was scanned with a 1.5 μm resolution to quantify the defect size and distribution. Commercial software CTAn (Skyscan, Belgium) was used to analyze the defects.

RESULTS AND DISCUSSION

Most ceramics are very poor thermal conductors. It would be natural to think different cooling rates will result in different flaw structures and mechanical behavior. This study examined whether different cooling rates affected the flaw size and distribution. A radiograph and tomography slices in Figure 3 - Figure 5 showed the position and size of flaws in the three crowns. Not all the flaws were shown, but all observed flaws lie between the two green lines marked on the radiographs. Most voids were found on the top one-third portion of the porcelain. There could be some defects in other part of the crown, but they would be harder to detect because of the large attenuation of zirconia. The biggest voids were 220 μm, 162 μm, and 132 μm for crowns with the normal, fast, and slow rates respectively. The fast rate showed 10 locations having flaws, while the slow rate had 3, and the normal cooling rate had 6 locations. The crown with the fastest cooling rate had more locations with flaws, but the crown with the slower cooling rate had the largest flaw. Additional research on a larger number of samples would provide more conclusive statistics. Also, cooling rate effects on thermal properties of the crowns had not been investigated in this work and would be a future research interest.

To quantify the defect size and distribution, a high resolution scan, 1.5 μm, was performed on the top 2.85 mm thickness of the porcelain cusp in the slow cooled crown. The higher resolution scan produced an image with rich material information and large size digital data. Because of limitations of computation capacity, only a cube area with the size of 1.275x1.275x1.275 mm was considered for defect size and distribution, as shown in Figure 6. The equivalent sphere diameter was used as a parameter to compare the size distribution of defects. The results indicated that most defects were below 30 μm, and there were a couple larger than 50 μm. The higher resolution scan revealed defects at a smaller size which is not obviously seen at lower resolution as shown in Figure 3 to 5.

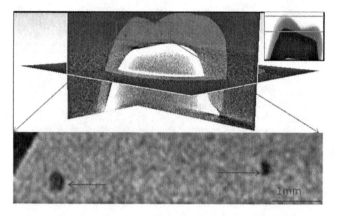

Figure 3 Tomography slices and radiograph (top right corner) of the normal cooling rate crown. Voids as big as 220 μm were observed. All the observed flaws were located between two green lines marked in the inset radiograph.

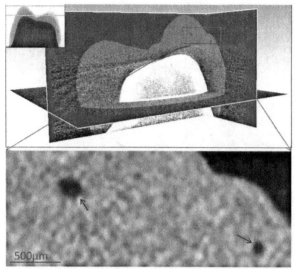

Figure 4 Tomography slices and radiograph (top right corner) of the fast cooling rate crown. Voids as big as 162 μm were observed. All the observed flaws were located between the two green lines marked in the inset radiograph.

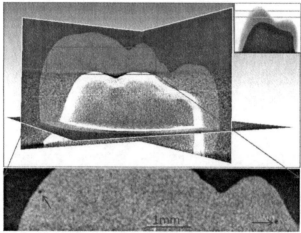

Figure 5 X-ray images from the slow cooled crown similar to the previous figure. Voids as large as 132 μm were observed. All the observed flaws were located between two green lines marked in the inset radiograph.

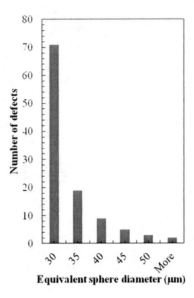

Equivalent sphere diameter (μm)

Figure 6 Porcelain defect size distribution in a slow cooled crown scanned at high resolution.

Both the number and size of flaws will affect the fracture stress. If the flaw size was the dominant factor for fracture, the critical fracture stress could be predicted based on Griffith Mode I fracture [17]

$$\sigma_c = \frac{K_{Ic}}{\varPsi c^{1/2}} \tag{2}$$

Where, σ_c and K_{Ic} are the critical fracture stress and critical stress intensity factor, or fracture toughness, $\varPsi$ is a crack geometry related factor, and c is the critical crack length. The flaws were treated as penny-shaped cracks. Porcelain fracture toughness reported in the literature [6;18] are from 0.6~1.5 (MPa.m$^{1/2}$). The value use here was 0.7 MPa.m$^{1/2}$, measured previously on another crown sample. The critical flaws were 220 μm, 162 μm, and 132 μm for the normal, fast, and slow cooled crowns. The predicted critical stress worked out to 42, 49, and 54 MPa, if the crown fails due to the size of its biggest flaws. Typical stresses on the crown in service are caused by a range of geometries. Typical loads can range from 30 to 530 N [19-20]. Assuming a contact area of 1 mm^2 the above stresses suggested failures at loads as low as 42 N! Using this μCT method, experimental validation could be employed to determine the true load to failure and lifetimes associated with flaws and establish a maximum flaw size or shape safe for use in dentistry.

CONCLUSION

Microtomography was used to evaluate the number and size of the flaws in three crowns made at different cooling rates. The flaws were clearly observed in the tomographs, with the biggest flaw observed at 220 μm. A simplified view suggests such crowns can fail at a stress of 42 MPa, if the flaw size was the dominant factor. The cooling rate seemed to affect the number and size of flaws. Additional research on larger number of samples is needed.

ACKNOWLEDGEMENTS

The authors appreciate Richard Kelly of Nautilus Dental for providing the samples. Portions of this research were made possible by an Oklahoma Health Research award (project number HR07-134), from the Oklahoma Center for the Advancement of Science and Technology (OCAST) at the Helmerich Advanced Technology Research Center.

REFERENCE

1. Lawn, BR, Deng, Y, Lloyd, IK, Jananl, MN, Rekow, ED, Thompson, VP, "Materials design of ceramic-based layer structures for crowns", *Journal of Dental Research*, 81, pp 433-438 (2002)

2. Lorenzoni, FC, Martins, LM, Silva, NRFA, Coelho, PG, Guess, PC, Bonfante, EA, Thompsaon, VP, Bonfante, G, "Fatigue life and failure modes of crowns systems with a modified framework design", *Journal of Dentistry*, 38, pp 626-634 (2010)

3. Kelly, JR, Denry, I, "Stabilized zirconia as a structural ceramic: An overview", *Dental Materials*, 24 , pp 289-298 (2008)

4. Lohbauer, U, Amberger, G, Quinn, GD, Scherrer, SS, "Fractographic analysis of a dental zirconia framework: a case study on design issues", *Journal of the Mechanical Behavior of Biomedical Materials*, 3, pp 623-629 (2009)

5. Gonzaga, CC, Cesar, PF, Miranda, Jr. WG, Yoshimura, HN, "Slow crack growth and reliability of dental ceramics", *Dental Materials*, 27, pp394-406 (2011)

6. Taskonak, B, Griggs, JA, Mecholsky, Jr JJ, Yan, J, "Analysis of subcritical crack growth in dental ceramics using fracture mechanics and fractography", *Dental Materials*, 24, pp700-707 (2008)

7. Miranda, P, Pajares, A, Guiberteau, F, Cumbrera, FL, Lawn, BR, "Role of flaw statistics in contact fracture of brittle coatings", *Acta materialia*, 49, pp3719-3726, (2001)

8. Mainjot, A.K., Schajer, G.S., Vanheusden, A.J., Sadoun, M.J., "Residual stress measurement in veneering ceramic by hole-drilling", *Dental Materials*, 27, pp439-444 (2011)

9. Zhang, Y, Allahkarami, M, Hanan, JC. "Measuring residual stresss in ceramic zirconia-porcelain dental crowns by nanoindentation", *Journal of Mechanical Behavior of Biomedical Materials*, 6, pp120-127 (2012)

10. Hounsfield, GN, "Computerized transverse axial scanning (tomography): Part 1, Description of system", *The British Journal of Radiology* 46 (552), pp 1016–1022, (1973)

11. Long, H, Swennen, R, Foubert, A, Dierick, M, Jacobs, P, "3D quantification of mineral components and porosity distribution in Westphalian C sandstone by microfocus X-ray computed tomography", *Sedimentary Geology*, 220, pp 116–125, 2009

12. Ferreira de Paiva, R, 1995. "De´veloppement d'un microtomographe X et application a` la characte´risation des roches re´servoirs", *The`se de doctorat de l'Universite´ de Paris* VI, p170 (1995)

13. Yao, Y, Liu, D, Che, Y, Tang, D, Tang, S, Huang, W, "Non-destructive characterization of coal samples from China using microfocus X-ray computed tomography", *International Journal of Coal Geology*, 80, pp 113–123(2009)

14. Van Geet, M, Swennen, R, Wevers, M, "Quantitative analysis of reservoir rocks by microfocus X-ray computerised tomography", *Sedimentary Geology*, 132, pp 25–36, (2000)

15. Hanan, JC, Veazey, C, Demetriou, MD, DeCarlo, F, Thompson, JS, "Microtomography of amorphous metal during thermo-plastic foaming", *Adv. X-Ray Anal.*, 49, pp 85-91 (2006)

16. Cnudde, V, Cnudde, JP, Dupuis, C, Jacobs, PJS, "X-ray micro-CT used for the localization of water repellents and consolidants inside natural building stones", *Materials Characterization*, 53, pp 259– 271(2004)

17. Lawn, Brian R., "Fracture of Brittle Solids, 2nd edition" *Cambridge University Press*, 1993

18. Morena, R, Lockwood, PE, Fairhurst CW, "Fracture toughness of commercial dental porcelain" *Dental Materials*, 2, pp 58-62 (1986)

19. Fernandes, CP, Glantz, PJ, Svensson SA, Bergmark A. "A novel sensor for bite force determination", *Dental Mateials*, 19, pp118-126, (2003)

20. Paphangkorakit, J and Osborn, JW. "Effects on human maximum bite force of bitting on a softer or harder object", *Archives of Oral Biology*, 43, pp833-839 (1998)

RESIDUAL STRESS AND PHASE TRANSFORMATION IN ZIRCONIA RESTORATION CERAMICS

M. Allahkarami[1], J. C. Hanan[1*]

[1]Mechanical and Aerospace Engineering, Oklahoma State University

Tulsa, OK, USA

ABSTRACT

A ceramic based dental restoration's service life is generally limited to a few years until catastrophic fatigue cracking. Materials improvements over the last few years have helped extend performance. With patient life expectancy continuing to increase, and limits on available biocompatible material options, the need for more understanding of existing materials used in crown design continuous to grow. Future generations of multi layer ceramic restorations can benefit from the manufacture process and geometries to better control residual stress at the contact surface and internal interfaces. As an example, having compressive residual stress beneath applied external loads increases damage resistance and delays crack initiation. Measurement techniques interpreting internal stress states are in development. At different locations of the crown, residual stresses as large as ±400 MPa and tetragonal to monoclinic phase transformations on a typical zirconia-porcelain crown was measured using a laboratory micro-diffraction system.

INTRODUCTION

The aptitude of ceramics to be color matched with adjacent teeth provides an excellent esthetic quality for these materials when replacing metal alloy restorations. However, all-porcelain dental restorations are not strong enough for most applications. The solution was to include a core of alumina or zirconia ceramic veneered by porcelain. This complicates fabrication as porcelains fire at lower temperatures than typical engineering ceramics. Other solutions have also developed, but most today include zirconia. Stabilized polycrystalline tetragonal zirconia has high fracture toughness and crack resistance for a ceramic. Despite considerable improvement in the mechanical properties of all ceramics dental crowns, there remain cases of catastrophic failure of these materials in service [1-2]. Chipping and crack propagation in the porcelain veneer or along the interface at the porcelain side are the most common failure modes in zirconia-porcelain bi-layer systems [3-4]. Residual stresses in ceramic bi layer ceramic systems are unavoidable. They can be attributed to the sintering process, thermal expiation mismatch, and an elastic modules difference between the layers [5]. The magnitude of residual stress depends on the preparation process and design parameters like layer thicknesses [6]. Polycrystalline tetragonal zirconia may transform to its monoclinic phase under a critical stress level and release the stored stress, called stress relaxation phase transformation [7]. This stress releasing phase transformation delays failure. When a catastrophic crack starting from the porcelain layer reaches the core layer, tetragonal zirconia initially transforms to monoclinic and resists against fracture [8]. H. A bale, *et al.* used highly focused monochromatic synchrotron X-rays to provide information on the biaxial residual stress in zirconia crown and observed stresses between -400 to 100 MPa along a 500 µm line scan [9], they have also reported much larger stresses at smaller length scales even up to ± 1GPa [10,11].

---

* Corresponding author; e-mail: jay.hanan@okstate.edu; phone: 918-594-8238.

Cooling rate, annealing, composition and also design parameters like layer thicknesses can affect the residual stress, so it is advantageous to adapt laboratory methods like X-ray diffraction [12, 13] or nano indentation [14,15,16] to study residual stress in a large variety of dental crowns provided by different venders. In this paper we will explain a method that utilizes a Bruker D8 Discover XRD[2] micro-diffraction system equipped with a Hi-star 2D area detector to collect and analyze the residual stress on the zirconia side of a typical failed zirconia-porcelain bi-layer dental crown.

MATERIALS AND METHODS

Molar crowns samples with a 0.5 mm thickness core were provided by NYU Dental School. The ceramic core was yttrium stabilized tetragonal polycrystalline zirconia machined by precise computer numerical control (CNC) milling and veneered by porcelain. A customized load frame was implemented to apply load ($\approx$500 N) for fracture by a 0.95 mm radius tungsten carbide indenter. Due to failure at one of the cusps, the porcelain veneer layer chipped off. Figure 1a, shows an optical image of the crown after it was sectioned into halves using a low speed water cooled diamond saw and mounted onto an aluminum plate with a high strength epoxy. Polishing was not implemented to avoid potential damage to the sections.

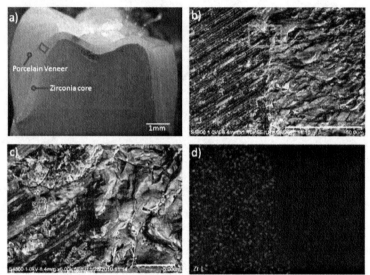

Figure 1a) An optical microscope image of sectioned sample. b) SEM micrographs of interface c) zoom image d) Edax element mapping shows the distribution of zirconium element at interface of zirconia and porcelain.

Figure 1b illustrates a scanning electron microscopy (SEM) image of the zirconia-porcelain interface. Cutting marks left behind by the diamond cutter are visible just on the zirconia side. While in low magnification bonding between porcelain and zirconia appears uninterrupted, a higher magnification image of the interface exposes small micro cracks less than 5 µm running along the interface. Edax element mapping illustrated in Figure 1d shows the distribution of zirconium element at interface region of zirconia-porcelain.

As all XRD stress measurement methods are based on peak location shifts, determining the exact peak position is an important step and requires adequate peak fitting. A proper peak fitting requires high quality data. Collecting the frames at a high 2θ angle with a 50 µm collimator by a laboratory X-ray system requires long exposure times, on the order of an hour per frame, because the total intensity that reaches the detector is as low as 25 counts per second. Collecting several numbers of these frames requires finding a reasonable time per frame. A set of frames with times from 1 minute to 120 minutes were collected. Results frames and selected integrated regions are illustrated in Figure 2a, and b respectively. A Matlab code was developed to fit a combination of two Pearson VII functions [17] on the (004) and (220) peaks as shown in Figure 2c. The curve $y(2\theta)$ is the intensity as a function of angular position 2θ that we want to fit on measured intensity $I(2\theta)$ around the (004) and (200) peaks. Because the (004) and (200) peaks overlap, a combination of two functions each represent one of the peaks.

$$y(2\theta) = A_1 f_1 (2\theta - 2\theta_1) + A_2 f_2 (2\theta - 2\theta_2) \tag{1-1}$$

Where A_1, A_2, θ_1 and θ_2 are intensity of fitted peaks and angular position of fitted peaks centers, respectively.

The f_1 and f_2 are Pearson VII distribution functions [18] defined as:

$$f_i (2\theta - 2\theta_i) = [1 + K_i^2 (2\theta - 2\theta_i)^2]^{-M_i} \tag{1-2}$$

Where K_i determines the width of the fitted curve and M_i governs the rate of decay of the tails. For $M_i = 1$, the profile is purely Cauchy, for $M_i = 2$, a Lorentzian function, and for M_i=infinity, the profile is purely Gaussian.

Although increasing the exposure time per frame provides less noisy data, there is an experimental upper limit for collection time after which the ratio of the signal to noise remains constant. The Mean Square Error (MSE) of the data from the fitted curves that represents the signal to noise ratio is plotted in Figure 3. Based on this, a 45 minute exposure was selected per frame.

Peaks related to two different phases were identified using DIFFRACplus EVA search and match software linked to the ICDD (International Center for Diffraction Data) phase identification cards.

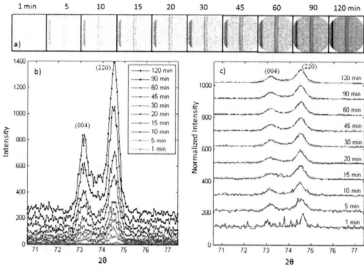

Figure 2 a) Frames with various times b) integrated pattern c) normalized and fitted peaks by two Pearson VII

functions

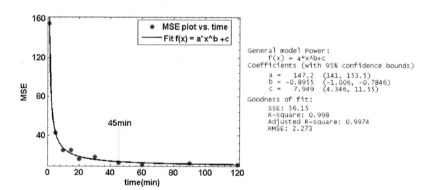

Figure 3 Mean square error of curve fits on (004) and (220) peaks saturates by 45 min.

RESULTS AND DISCUSSION

Residual stress and phase transformation measurements were done using the Bruker D8 discover XRD[2] micro-diffraction system equipped with a Hi-star 2D area detector [19]. Bruker's General Area Diffraction Detection System (GADDS) was used for data collection. The maximum possible detector to sample distance (299.5 mm) by this system was selected. Frames collected at a long detector distance have more strain resolution and diffraction rings are more distinguishable, although diffraction ring intensity decreases. At 300 mm the detector simultaneously covers the field of view of 20° in 2θ and 20° in χ with a 0.02° resolution. The 2θ interval of 63° to 83° at each diffraction point (X,Y) was considered for simultaneous $\sin^2\psi$ bi axial stress measurements and phase transformation. Frames with a 45 minute exposure were collected using Cu-Kα radiation at tube parameters of 40kV/40mA with a 50 μm diameter collimator in reflection mode. A motorized five axis (X, Y, Z, χ (tilt), φ (rotation)) sample stage was used for precise sample movement. Angles are shown schematically in Figure 4. Height adjustments in the Z-direction were made by an auto video-laser system before each frame. Prior to conducting the experiment, calibration of the system was checked by collecting a diffraction pattern from a NIST standard polycrystalline corundum sample.

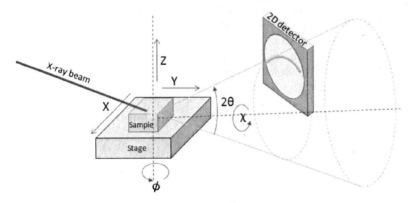

Figure 4 Schematic of the sample orientation and position of the 2d detector with respect to the diffraction cones. χ is the angle subtended by the diffraction rings on the 2D detector.

The applied stress was concentrated in small area underneath the ball indenter tip, a good candidate location for phase and stress analysis is a line scan right under the indenter load. Figure 5b shows the location of exposure points selected on a line that connects the bottom and top side of the zircona layer. Frames collected from points 1 and 23 illustrated in Figure 5 c and d are substantially different. This difference indicates stress induced change in crystal structure.

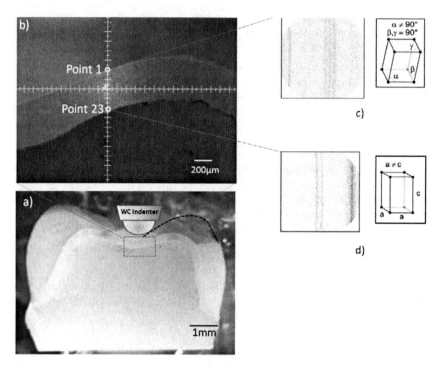

Figure 5 a) An optical microscopy image of the measurement location on the crown b) exposure points, starting at the interface between zirconia and porcelain (point 1) and proceeding across the zirconia layer away from the interface (point 23) c) Monoclinic 2-D diffraction image (2θ=63° to 83°) d) Tetragonal 2-D diffraction image (2θ=63° to 83°).

Observation of phase transformation from X-ray diffraction frames

A mosaic pattern of frames collected along the cross section line that connects the surfaces of the zirconia layer under the indentation point qualitatively reveals phase transformation, as shown Figure 6 . Frames collected at locations 1 and 23 have different sets of rings identifying the two different phase of zirconia, but frames at the middle of the length (frame number 12) show both phases.

Figure 7 illustrates χ integration of 2D diffraction data as intensity versus 2θ for all 23 frames. Peaks observed from the frame collected at the first point matches with 01-070-4426, tetragonal Zirconium Yttrium Oxide. Peaks observed at the frame collected at point 23 matches with 01-089-9066, monoclinic Zirconium Yttrium Oxide. As illustrated in Figure 7 diffraction patterns across the thickness for all measurement points reveal a gradual phase transition from tetragonal to less symmetric monoclinic crystal structure. Scanning a similar line on a sectioned crown without impact did not show monoclinic phase. This reveals that the phase transition discussed above was created by impact damage. Observation of this impact related phase change suggests the hypothesis that the veneered layer failed finally as a result of external load, but the zirconia layer did not fail and transformed from tetragonal to monoclinic. In this experiment, the sequence of the frames that gradually show phase transformation were collected. These frames were used to investigate the relation between the final stress state and the degree of tetragonal to monoclinic phase transformation in zirconia.

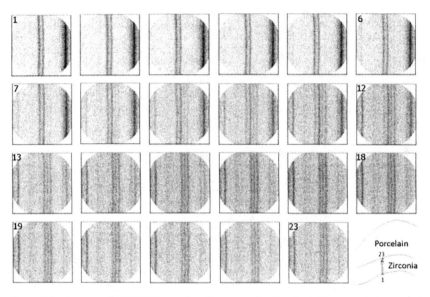

Figure 6 Mosaic pattern of 2D XRD frames along the cross section from the inside surface to the porcelain interface that cover the $\chi = 0°$ to $20°$ segment of diffraction cones

The ratio of area under three monoclinic peaks to the tetragonal peaks was used as an indication of phase transformation percentage. For the monoclinic phase, the sum of areas under the (-231), (140), (-411) peaks, and for the tetragonal phase the area under the (004), (220) and (-411) peaks were considered. The result of this phase transformation percentage is shown in Figure 8 along with the result of the residual stress measurement. For residual stress, the integrated intensity for $1°$ χ segments of collected frames were fit by two Pearson VII functions using a Matlab program. The (004) tetragonal zirconia peak at $73.065°$, and (231) monoclinic zirconia peak was selected for $\sin^2 (\chi)$ measurement. Details of the method for residual stress measurement have been published elsewhere [8]. The error bars in measured stresses were determined based on the confidence bound of the peak fitting and variance of seven measurements at each location. The calculated confidence bound corresponds to a 2θ range that itself corresponds to a range in d-spacing. The biaxial residual stress was found to range from -400 MPa to 400 MPa for the scanned line. An important point to note is initially at point 1 the stress was -400 MPa (compressive) and gradually increased to +400 MPa (tensile) at point 7, and then dropped to zero (average) after this peak. The gradual changes in the phase transformation ratio with this gradual stress change from compression to tensile, then to zero, revels a residual stress relaxation from phase transformation. This mechanism of phase transformation prevents the zirconia layer from crack growth in its application as a dental material combined with a porcelain veneer. Considering the fact that meta stable tetragonal zirconia transforms to monoclinic due to tensile stress, and the current observed relation between stress and phase ratio, determines an important role for compressive residual stress at the zirconia layer interface.

Before applying any external load to a typical zirconia-porcelain crown, depending on cooling process, geometry and relative thickness of layers, a residual stresses from -100 MPa to +100 MPa is expected along the thickness of zirconia layer [6]. If the zirconia layer is in compression before applying external tensile load, initially the compressive residual stress might be compensated and only further tensile loading would result in phase transformation. It is here that the meta-stable tetragonal zirconia relaxes the tensile stress by performing phase transformation and likely delays crack initiation.

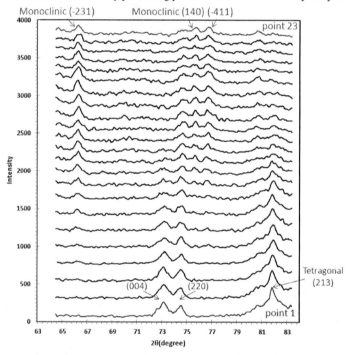

Figure 7 Intensity versus 2θ diagrams across the thickness, from integrated rings in χ direction.

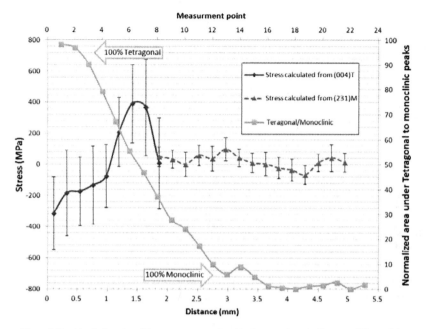

Figure 8 Combined phase transition percentage and residual stress along the line scan (Figure 4b)

The intercept of the $\sin^2\chi$ versus χ curve was used as reference for residual stress measurements [12]. It was observed from Figure 8 that stresses go to zero at about 50% phase transformation. The current measurement was performed in one line scan on a (004) peak and focusing mostly on introducing the technique. As future work, more measurements at other locations of the crown and considering other peaks would be beneficial. However, the technique is powerful and these first results are sufficient to indicate some interesting behavior. The transformation to a majority monoclinic phase 1.5 mm from the interface would also mean communication of stresses from gran-to-grain within this tetragonal phase becomes limited. Thus the stress in the monoclinic phase would dominate in this region. The transformation to monoclinic during impact would have been associated with a volume expansion. Once the driving force from the applied load was released, the residual stress state would have changed. It is possible that this volume change is also associated with micro-cracking. Future work could also include microscopic evaluation to observe cracking. That micro-cracking would have limited residual stresses in this part of the sample and supports the current observation of nominally 0 MPa stress beyond 1.8 mm from the interface. Further future work could also include observing the stresses in the direction parallel to the interface. The stresses examined here were aligned primarily perpendicular to the interface. In addition, implementing this method on a specimen fatigued to failure would be clinically relevant.

CONCLUSIONS

The presence of both monoclinic and tetragonal phases was confirmed using a laboratory X-ray diffraction system. A method of combined phase transformation and residual stress measurement from 2D X-ray diffraction data was adapted to study zirconia core dental crowns. It was observed that polycrystalline meta stable tetragonal zirconia transforms up to 100% to its less symmetric monoclinic phase as a result of external load and failure. A ±400MPa residual stress interval was measured for an impacted crown under a single load to failure. This study implements an experimental method to map the phase transformation, after applying local compressive load until fracture. Such fractures resemble clinically observed chipping failure. Understanding the residual stress state and phase transformation in a crown and implementing controlled stresses for a new generation of crowns provides another tool for producing high performance, reliable dental restorations.

ACKNOWLEDGEMENTS

The authors would like to appreciate Dr. Hrishikesh Bale for his work in indentation and Drs. P. Coelho, V. Thompson, E. D. Rekow and M. Cabrera at NYU College of Dentistry for providing the samples. Partial research support provided by the Oklahoma Health Research award (project number HR07-134), from the Oklahoma Center for the Advancement of Science and Technology (OCAST) and the Helmerich Advanced Technology Research Center, Oklahoma State University, Tulsa.

REFERENCES

[1] B. Kim, Y. Zhang, M. Pines, V.P. Thompson, Fracture of porcelain veneered structures in fatigue. *J Dent Res*, **86**, 142–146, (2007)

[2] J.W. Kim, J.H. Kim, V. Thompson, Y.Zhang, Sliding contact damage in layered ceramic structures, *J Dent Res*, ,**86** , 1046–1450, (2007)

[3] J.W. Kim, J.H. Kim, M.N. Janal, Y. Zhang, Damage maps of veneered zirconia under simulated mastication, *J Dent Res*, **87** (12), 1127–1132, (2008).

[4] I. Sailer, A. Feher, F. Filser, L.J. Gauckler, H. Luthy, C.H. Hammerle, Five-year clinical results of zirconia frameworks for posterior fixed partial dentures. *Int J Prosthodont*, **20**, 383–388, (2007)

[5] S. Timoshenko, Analysis of Bi-Metal Thermostats, *J. Opt. Soc. Amm*, **11**, 233–255 (1925)

[6] M. Allahkarami, H. A. Bale, J. C. Hanan, Analytical model for prediction of residual stress in zirconia porcelain bi-layer, *Advances in Bioceramics and Porous Ceramics III: Ceramic Engineering and Science Proceedings, John Wiley & Sons, Inc*, **31**, 19-26, (2010)

[7] M. Allahkarami, J. C. Hanan, Mapping the tetragonal to monoclinic phase transformation in zirconia core dental crowns, *Journal of Dental materials*, **27**(12), 1279-1284, (2011)

[8] M. Allahkarami, J. C. Hanan, Residual Stress Delaying Phase Transformation in Y-TZP Bio-restorations, Phase Transitions, **85** (2011),

[9] H. A. Bale, J. C. Hanan, Microbeam X-ray grain averaged residual stress in dental ceramics, *Advances in Bioceramics and Porous Ceramics IV: Ceramic Engineering and Science Proceedings*, **32**, 49-63 (2011)

[10] H. A. Bale, J. C. Hanan, N. Tamura, Average and Grain Specific Strain Resolved in Many Grains of a Composite Using Polychromatic Microbeam X-Rays. *Adv. X-Ray Anal*, **49**, 369-374, (2006)

[11] H. A. Bale, N. Tamura, P. Coelho, J. C. Hanan, Interface Residual Stresses in Dental Zirconia Using Laue Micro-Diffraction, *Advances in X-ray Analysis*, **52** (2008).

[12] A. D. Krawitz, Introduction to diffraction in materials, science, and engineering, chapter 10, *John Wiley*, (2001)

[13] P. S. Prevey, X-ray Diffraction Residual Stress Techniques, *Metals Handbook, Metals Park, American Society for Metals*, **10**, 380–392 (1986)

[14] S. Suresh, A. E. Giannakopoulos, A new method for estimating residual stress by instrumented sharp indentation Acta Mater, **46**(16), 5755-5767 (1988)

[15] Y. Zhang, J. C. Hanan, Residual stress in ceramic zirconia-porcelain crowns by nanoindentation", *Mechanical Properties and Performance of Engineering Ceramics and Composites VI*, 67-76, (2011)

[16] Y. Zhang, M. Allahkarami, J. C. Hanan, Measuring residual stress in ceramic zirconia-porcelain dental crowns by nanoindentation, *Journal of the Mechanical Behavior of Biomedical Materials*, **6**, 120-127, (2011)

[17] M.T. Hutchings, P.J. Withers, T.M. Holden, T. Lorentzen, Introduction to the Characterization of Residual Stress by Neutron Diffraction , *Chapter 4, CRC press, Taylor & Francis*, (2005)

[18] P. S. Prevéy, The use of pearson vii distribution functions in X-ray diffraction residual stress measurement, *Advances in X-Ray Analysis*, **29**, 103-111, (1986)

[19] B. B. He, Two-Dimensional X-ray Diffraction, *Hoboken, NJ, USA, John Wiley and Sons*, (2009)

HETEROGENEOUS STRUCTURE OF HYDROXYAPATITE AND IN VITRO DEGRADABILITY

Satoshi Hayakawa, Yuki Shirosaki and Akiyoshi Osaka
Graduate School of Natural Science and Technology, Okayama University
Okayama, Japan

Christian Jäger
BAM Federal Institute for Materials Research and Testing
Berlin, Germany

ABSTRACT

Hydroxyapatite (HAp) particles were synthesized by two processing methods such as solid-state reaction and wet chemical reaction, and were characterized in terms of their chemical composition, disordered structure and in vitro biodegradability.   XRD revealed that the prepared HAp particles were composed of single phase HAp, while 1D and 2D solid-state NMR analysis showed that the prepared HAp particles consisted of not only crystalline HAp but also disordered phase.   The in vitro biodegradability was discussed by using a structure model for nano-crystalline HAp, in which the nano-crystals consist of a crystalline HAp core covered with a disordered surface layer (core-shell model).

INTRODUCTION

Stoichiometric hydroxyapatite (pure HAp) is chemically and thermodynamically stable than non-stoichiometric HAp.[1]   For example, commercially available synthetic HAp bioceramic materials for clinical use are non-degradable and non-resorbable.   On the other hand, in bone tissue, HAp is non-stoichiometric and thus the non-stoichiometric HAp with Ca/P ratio of less than 1.67 exhibits the high biodegradability.   Since HAp is decomposed by thermal treatment above ca.1200°C, accompanied by forming partially dehydrated oxyhydroxyapatite (OHAp), or forming tricalcium phosphate (α-TCP), depending on the calcium deficiency, it is possible that heat-treated HAp has a disordered structure.[2]   However, it is not clear how the non-stoichiometry and the disordered structure of HAp influence the biodegradability.

In this study, HAp particles were prepared by two processing methods such as conventional high-temperature solid-state reaction or wet chemical reaction, and were characterized in terms of in vitro biodegradability, as well as their structures by means of X-ray diffraction technique, Fourier-transform infra-red (FT-IR) and solid-state nuclear magnetic resonance (NMR) spectroscopies.   The disordered phase and local environment around phosphorus atoms were correlated to the in vitro biodegradation behaviors.

EXPERIMENTAL

(1) Materials and Methods

Hydroxyapatite (HAp) particles were synthesized by using reagent-grade calcium hydroxide ($Ca(OH)_2$) and orthophosphoric acid (H_3PO_4) through the wet chemical procedure[3], where the molar ratio Ca/P was ca.1.68 near stoichiometric HAp (Ca/P=1.67).   Precipitations were obtained by titrating 100 mL suspension (pH10.5) of $Ca(OH)_2$ (0.500M) simultaneously with 100 mL solution of H_3PO_4 (0.297M), held in a three-neck flask with a magnetic stirrer, at a dropwise rate of 3 mL/min at room temperature (RT) or 80°C.   After the completion of the dropwise addition, the precipitates were aged for 15 hours at RT, washed with distilled deionized water, and dried at 80°C for 48 hours.   Those processes were conducted in a glove box under flowing dry N_2 gas.   The derived cakes were milled and sieved to obtain HAp particles 300~600 μm in size.   They were denoted as HAp_x, where x stands for the reaction temperature (RT or 80).   Commercially available reagent-grade calcium carbonate ($CaCO_3$), and calcium pyrophosphate ($Ca_2P_2O_7$) prepared through calcining reagent-grade dicalcium phosphate dihydrate (DCPD) at 1000°C, were mixed well in an alumina mortar for 30 min and pelletized to $\phi 10$ mm×15 mm disks. They were placed in a platinum crucible in an electric furnace, and heated up to 1473 K at the rate of 5 K/min, and kept at 1200°C for 6 h. After quenching, the pellets were pulverized. The pelletizing and heating processes were iterated twice. The pellets were milled and sieved to obtain HAp particles 300~600 μm in size (denoted as HAp_solid). For comparison, commercially available HAp (HAP-100, $Ca_{10}(PO_4)_6(OH)_2$, Taihei Chemical Industry Ltd., Osaka, Japan) was used as a reference for characterization of structure and *in vitro* biodegradability. The chemical compositions of the prepared HAp particles were analyzed by inductively coupled plasma emission spectroscopy (ICP, iCAP6000, Thermo Fisher Scientific, Yokohama, Japan).

Table I summarizes the sample codes, chemical compositions, and atomic ratios of Ca/P for the HAp particles.

Sample code	Ca (mass%)	P (mass%)	Ca/P molar ratio
HAp_RT	41.2	19.6	1.63
HAp_80	39.8	18.1	1.70
HAp_solid	44.7	21.1	1.64
HAP-100	-	-	1.65

The crystalline phases were identified by an XRD (X'Pert-ProMPD, PANalytical, Almelo, the Netherlands; CuKα, 1.5418 Å, 45 kV, 40 mA). The line broadening of the (002) and (300) reflections was used to evaluate the length of the coherent domains (d_{hkl}, i.e., the crystallite size) along c-axis and a-axis using the Scherrer equation. The Fourier-transform infra-red (FT-IR) spectra were measured by the KBr method on a JASCO FT/IR-300 spectrometer (JASCO Co., Tokyo, Japan) with 300 scans and a resolution of 4 cm^{-1}.

The local structure around the phosphorus and protons in the HAp particles was examined by solid-state magic-angle spinning nuclear magnetic resonance spectroscopy (MAS-NMR). All the NMR experiments were carried out on a Varian ^{UNITY}INOVA300 FT-NMR (Fourier-transformed NMR) spectrometer (Varian, Inc., Palo Alto, CA, USA), equipped with a cross-polarization(CP)-MAS probe. A silicon-nitride rotor with a diameter of 7 mm was used for ^{31}P and ^{1}H NMR measurements. The rotor spinning frequency was 5.0 kHz. Two-dimensional (2D) ^{1}H→^{31}P heteronuclear correlation (HETCOR) experiments were performed using cross-polarization with contact time of 1 ms, recycle delay of 5 s and 100 scans per t_1 increments; 108 t_1 slices were acquired, where the rotor spinning frequency was controlled to be 5.0 kHz.[4,5] Direct polarization ^{31}P MAS-NMR spectra were taken at 121.4 MHz where a 3.3-μs pulse length (π/4-pulse angle) and 90-second recycle delays. The signals from about 40 pulses were accumulated with 85% H_3PO_4 as the external reference (0 ppm). ^{1}H high-power decoupling was used during the ^{31}P acquisition. ^{1}H MAS-NMR spectra were taken at 299 MHz with a 3.45-μs pulse length (pulse angle, π/2) and 5-second recycle delays, where the signals from about 40 pulses were accumulated with tetramethylsilane (TMS) as the external reference (0 ppm).

N$_2$ adsorption isotherms for dried HAp particles were obtained by multi-point N$_2$ gas adsorption experiments (Micrometrics GEMINI2370, SHIMADZU, Kyoto, Japan). The specific surface area (SSA) was derived by the BET N$_2$ adsorption method.

(2) In Vitro Biodegradation Experiment

The in vitro degradation behaviors of the HAp particles was evaluated with the change of the Ca(II) concentration in an acetic acid and sodium acetate (AcOH–AcONa) buffer solution as a function of time, according to the method developed by Ito *et al.*[6] The pH of the buffer model (pH5.5, 25°C) would simulate osteoclastic resorption.[7] Pulverized HAp particles (10 mg, sieved to 300-600 μm in size for HAp_solid, 45-300 μm in size for HAp_RT, HAp_80 and HAP-100) were soaked in 100 mL of the AcOH–AcONa buffer solution at pH5.5 under stirring at a rate of 450 rpm. The calcium concentration was continuously measured with a pH/ion meter (F-53, Horiba, Kyoto, Japan) equipped with a Ca ion electrode (6583-10C, Horiba). All experiments were performed in triplicate.

RESULTS AND DISCUSSION

ICP analysis results (Table I) indicated that all HAp particles had 1.63-1.65 or 1.70 of Ca/P ratios, were non-stoichiometric. The whole chemical composition and the chemical formulas of the samples cannot be determined only by ICP analysis, because all HAp particles were calcium-deficient below Ca/P of 1.67 (HAp_RT, HAp_solid, HAP-100) or calcium-rich (Ca/P=1.70, HAp_80) and may be accompanied by disordered phases.

Powder XRD patterns of the HAp particles are shown in Fig.1. All HAp particles gave XRD profiles, being similar to those of HAp crystal of space group P63/m (JCPDS#09-0432). It is noted that higher synthesis or heating temperature gave stronger and sharper XRD peaks of apatite, indicating that the crystallinity of HAp particles increased with the synthesis or heating temperature. The estimated crystallite size (CS) for 002 and 300 diffractions (Table II) showed that prepared HAp particles exhibited the similar anisotropic HAp crystal growth feature of CS ratio of ca.2. The values of specific surface area (SSA) also decreased with increasing the synthesis or heating temperature, being consistent with the crystallite size.

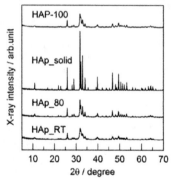

Figure 1   Powder XRD patterns of the HAp samples.

Table II   Crystallite size (CS) and specific surface area (SSA) of the HAp particles.

Sample code	Crystallite size (nm)		CS ratio	SSA
	002	300	002/300	m²/g
HAp_RT	39	15	2.6	104
HAp_80	70	35	2.0	52
HAp_solid	221	142	1.6	1.8
HAP-100	47	23	2.0	66

FT-IR spectra of the HAp particles in Figure 2 show the peaks characteristic of PO_4^{3-} ions at 565, 603, 962, 1034 and 1093 cm^{-1}, assignable to $v_4(PO_4)$, $v_1(PO_4)$ and $v_3(PO_4)$, respectively. In addition, the librational and stretching modes of OH⁻ in the apatite lattice appeared at ca. 631 cm⁻¹ and 3570 cm⁻¹, respectively, which accompanied a strong and broad band at around 3437 cm⁻¹ for water adsorbed on the surface.   As can be seen in Fig. 2, the 631 cm⁻¹ peak from librational mode of OH⁻ in the apatite lattice and the 3437 cm⁻¹ peak for water adsorbed on the surface became small by high temperature heating process (see HAp_solid).

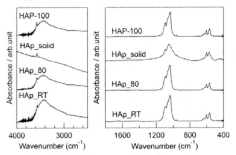

Figure 2   FT-IR spectra of the HAp particles.

The local structures around proton and phosphorus atoms were examined by solid-state NMR spectroscopy.   Figure 3 shows ¹H and ³¹P MAS-NMR spectra of the HAp particles.   Two resonances at 4.7-5.3 ppm in δ (¹H) and 0.0 ppm in δ (¹H) were characteristic of the H atoms in water molecules adsorbed on the surface and OH⁻ in the lattice of the HAp, respectively.   As obviously observed in Fig.3, the fraction of OH⁻ groups remarkably decreased by high temperature heating process (see HAp_solid), being in agreement with the FT-IR results.

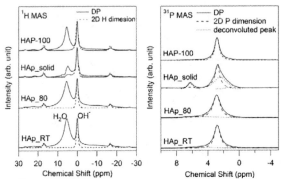

Figure 3   ¹H and ³¹P MAS-NMR spectra of the HAp particles.   Left: Slices taken at 2.8 ppm in the ³¹P dimension. The ¹H line shape for 2.8 ppm-slice (in δ(³¹P)) is mainly derived from OH⁻ in the HAp lattice.[2,5]   Right: Slices taken at 0 ppm in the ¹H dimension. The ³¹P line shape for 0 ppm-slice (in δ(¹H)) is representative of crystalline HAp.[2,5]

³¹P MAS-NMR spectra of the HAp particles (Fig.3) indicated two differences in their profiles as a width of a singlet peak at 2.8 ppm in δ(³¹P) assignable to PO_4^{3-} inside of HAp lattice structure and the intensity of the shoulder around 5.0-6.0 ppm in δ(³¹P).   HAp_solid indicated the asymmetric broadest ³¹P MAS NMR spectra.   This means greatest chemical distribution in the P(V) environment due to the formation of a disordered phase.   Two-dimensional (2D) ¹H→³¹P

HETCOR spectrum for HAp_solid particles (Fig. 4) shows an expected correlation peak between the intense ^{31}P signal at 2.8 ppm and the ^{1}H signal at 0 ppm characteristic of HAp structure, and the spectral profile of slice taken at 0 ppm in ^{1}H dimension was consistent with that of other HAp particles (see Fig. 3). In addition, it is noted that the weak ^{1}H signal at 4.7 ppm assignable to water molecules adsorbed on the surface correlates with a broad ^{31}P signal at 2.3 ppm. The weak 4.7 ppm/2.3 ppm and 7.7 ppm/1.4 ppm correlation peaks can be explained by considering that the phosphate ions (PO_4^{3-}) and hydrogen phosphate ions (HPO_4^{2-}) are located near water molecules adsorbed on the HAp_solid surface. Thus, we can assume that at least four kinds of phosphate species were present in the HAp_solid (peaks at 1.4, 2.3, 2.8 and 5.8 ppm in $\delta(^{31}P)$) for the quantitative analysis for Fig.3. The ^{31}P MAS NMR spectra of the HAp particles can be deconvoluted into at least two peaks at 2.8 ppm and ca.5-6 ppm by a least squares fitting algorithm.

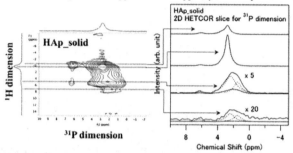

Figure 4 Left: 2D ^{1}H→^{31}P HETCOR spectrum with a contact time of 1 ms for HAp_solid. Five correlation peaks are observed: the intense peak at 0.2 ppm/2.8 ppm (^{1}H vs. ^{31}P shift) for crystalline HAp, -0.6 ppm/6.2 ppm and the weak 4.7 ppm/2.3 ppm, 7.7 ppm/1.4 ppm correlation peaks.
Right: Four slices taken at -0.6, 0.2, 4.7 and 7.7 ppm in the ^{1}H dimension. The ^{31}P line shape for 0.2 ppm-slice is representative of crystalline HAp.[2,5] The ^{31}P NMR peak deconvolution results, including peak position, FWHM and relative peak area (%) are listed in Table III.

Table III ^{31}P chemical shift (δ (ppm)) and full width at half maximum (FWHM (ppm)) of deconvoluted peaks derived from Fig 3.

Sample	δ (ppm)	FWHM (ppm)	Area%
HAp_RT	2.8	1.0	90
	4.9	3.9	10
HAp_80	2.9	1.3	89
	5.0	4.1	11
HAp_solid	1.4	1.0	4
	2.3	2.0	45
	2.8	0.7	42
	6.2	0.8	9
HAP-100	2.8	0.8	98
	5.8	1.1	2

The weak 2.3 ppm peak and broad 5-6 ppm peak (Table III) can be assignable to phosphate ions (PO_4^{3-}) in disordered phase far from OH^- groups inside the HAp lattice structure and hydrogen phosphate ions (HPO_4^{2-}) located near H_2O (water adsorbed on the surface), respectively. It is noted that the fraction of 2.8 ppm peak decreased in the order: HAP-100 > HAp_RT > HAp_80 > HAp_solid, while the FWHM of the peaks increased in the order: HAp_solid < HAP-100 < HAp_RT < HAp_80. The ^{31}P MAS NMR peak at 2.8 and 5-6 ppm for HAp_80 is broader than that for HAp_RT and means greater chemical distribution in the P (V) environment. In HAp_solid, the highest fraction is for the disordered phase, while the XRD analysis paradoxically indicated that almost all P atoms were derived from crystalline apatitic phase, which gave very sharp diffractions in Fig. 1. Therefore, the remarkable broadening of ^{31}P peak at 2.3 ppm of HAp_solid could be derived from the dehydroxylation; oxyhydroxyapatite (OHAp), described as follows[5]:

$$Ca_{10}(PO_4)_6(OH)_2 \rightarrow Ca_{10}(PO_4)_6(OH)_{2-2x}O_x + xH_2O \tag{1}$$

Since the present HAp particles consisted of a few phases involving disordered phase, equilibrium solubility is not a proper measure of their biodegradability. Thus, the concentration of calcium dissolved into the AcOH–AcONa buffer solution from the particles should be used to represent the biodegradation that simulates osteoclastic resorption. Figure 5 shows Ca ion dissolution curves for HAp particles together with the previous results on β-TCP and α-TCP.[5] It is clearly seen that the rate of Ca ion dissolution increased in the order: HAp_solid < HAP-100 < HAp_80 ≤ HAp_RT. This means that the rate of Ca ion dissolution of HAp particles mainly depends on their specific surface area. In addition, the in vitro biodegradation behaviors of the HAp particles are represented as the calcium concentration divided by surface area of these HAp particles, in order to discuss the effect of disordered phase or local atomic environment in HAp lattice structure on the rate of Ca ion dissolution behaviors.

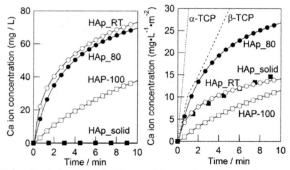

Figure 5   Ca ion dissolution curves for HAp particles together with those for β-TCP and α-TCP.

HAp_RT and HAp_80 released as many Ca ions as β-TCP in the initial stage, < 1 min, giving a parabolic profile. In the later stage, the rate of increase in the Ca concentration was very similar to that of HAP-100. As the core–shell model (Fig. 6) is applicable to both HAp_RT and HAp_80, it is strongly suggested that the core of HAp_RT and HAp_80 was almost pure HAp, while the shell was similar in chemical stability to β-TCP (Ca/P ratio of 1.5). Although the fraction of the shell (disordered phase, ca. 10%) of HAp_RT was almost the same with that of HAp_80, the rate of Ca ion dissolution of HAp_80 was distinctly higher than that of HAp_RT (ca. 2%). This result can be explained by considering that HAp_80 exhibited calcium-rich disordered surface layer, while HAp_RT had calcium-poor disordered surface layer and the difference in the local atomic order in the crystalline core of HAp between HAp_RT and HAp_80, giving greater chemical distribution in the P (V) environment. (Table III, FWHM of the peak at 2.8 ppm in $\delta(^{31}P)$; HAp_RT < HAp_80).

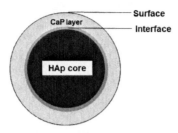

Figure 6   Core-shell type structural model for HAp having a crystalline core of stoichiometric HAp surrounded by a disordered layer of calcium phosphate (CaP).[2,4,5]

In contrast, HAp_solid was very similar to HAp_RT in the parabolic Ca ion concentration profile throughout the soaking period within 10 min, and the ratio of Ca/P in total was almost the same (ca. 1.64).   As the core–shell model (Fig. 6) is applicable to HAp_solid, the 2D $^1H{\rightarrow}^{31}P$ HETCOR NMR analysis and Table III indicated that HAp_solid is OHAp, in which crystalline core of HAp was surrounded by the disordered surface layer due to dehydroxylation.   Thus, it can be suggested that the disordered surface layer caused the enhanced release of Ca ions in the initial stage, < 1 min for HAp_RT, HAp_80 and HAp_solid.

CONCLUSION

An in vitro biodegradability test showed that wet chemically derived HAp particles degraded quicker than commercially available HAP-100.   Although main factor affecting the rate of Ca ion dissolution was the specific surface area, the disordered surface layer caused the enhanced release of Ca ions in the initial stage within 1 min, while the crystalline core of HAp also gave different release rate of Ca ions, depending on the chemical distribution in the P (V) environment (atomic disordering in the HAp lattice).

ACKNOWLEDGEMENT

This work was supported by a Grant-in-Aid for Scientific Research from the Japan Society for the Promotion of Science under the JSPS-DFG program and the Deutsche Forschungsgemeinschaft DFG under the JSPS-DFG program (grant no. Ja 552/25-1).

REFERENCES

[1] M. Vallet-Regi, D. Arcos, Biomimetic Nanoceramics in Clinical Use: From Materials to Applications, *RSC Publishing*, Cambridge, 2008.

[2] C. Jäger, T. Welze, W. Meyer-Zaika, and M. Epple, A solid-state NMR investigation of the structure of nanocrystalline hydroxyapatite, *Magn. Reson. Chem.*, **44**[6], 573–580 (2006).

[3] I.R. Gibson, S. M. Best, W. Bonfield, Chemical characterization of silicon-substituted hydroxyapatite, *J. Biomed. Mater. Res.* **44**[4], 422-428 (1999).

[4] S. Barheine, S. Hayakawa, A. Osaka, C. Jäger, Surface, Interface and Bulk Structure of Borate Containing Apatitic Biomaterials, *Chem. Mater.*, **21**[14], 3102-3109 (2009).

[5] S. Barheine, S. Hayakawa, C. Jäger, Y. Shorosaki, A. Osaka, Effect of Disordered Structure of Boron-Containing Calcium Phosphates on their *In Vitro* Biodegradability, *J. Am. Ceram. Soc.*, **94**[8], 2656-2662 (2011).

[6] A. Ito, K. Senda, Y. Sogo, A. Oyane, A. Yamazaki, and R. Z. LeGeros, 'Dissolution Rate of Zinc-Containing β-Tricalcium Phosphate Ceramics, *Biomed. Mater.*, **1** [3], 134–9 (2006).

[7]P. S. Eggli, W. Muller, and R. K. Schenk, "Porous Hydroxyapatite and Tricalcium Phosphate Cylinders with Two Different Pore Size Ranges Implanted in the Cancellous Bone of Rabbits. A Comparative Histomorphometric and Histologic Study of Bone Ingrowth and Implant Substitution, *Clin. Orthop. Relat. Res.*, **232**, 127–38 (1988).

ASPECTS OF ANTIBACTERIAL PROPERTIES OF NANOSTRUCTURAL CALCIUM
ALUMINATE BASED BIOMATERIALS

Leif Hermansson
Doxa AB
Uppsala, Sweden

ABSTRACT
This paper will give an overview of Ca-aluminate based biomaterials and the use within the
nanostructural biomaterials field. The paper describes typical characteristics of the Ca-aluminate
material with regard to chemistry and microstructure. Special focus will be on the microstructure,
which is in the nanosize range. The nanostructure including porosity of a few nm in size structures
opens up for some specific applications related to dental applications where antibacterial and
bacteriostatic aspects are of importance. Another field where nanosize porosity is essential is within
drug delivery systems for controlled release of medicaments among others antibiotics.

INTRODUCTION
Ceramics as biomaterials have been established within all the classical ceramic families;
traditional ceramics, special ceramics, glasses, glass-ceramics, coatings and chemically bonded
ceramics (CBC) [1]. Nanostructural biomaterials are commonly found within the CBC-materials.
Materials in this group are phosphates, aluminates, silicates and sulphates. Many of these materials
are known as inorganic cements, and have many application areas also outside those of
biomaterials, especially within the construction sector.

Most ceramics are formed at high temperatures through a sintering or a partial melting or a melting
process. By using chemical reactions, the biomaterials in the chemically bonded systems can be
produced at low temperatures (body temperature). This is attractive from several perspectives: cost,
avoidance of temperature gradients (thermal stress), dimensional stability and minimal negative
effect on the systems (tissues or other biomaterials) with which the material interacts. Notably,
apatite in hard tissues in humans is also formed via biological chemical reactions, and is close in
composition and nanostructure to those of the CBCs.

This paper will summarize important aspects of the Ca-aluminate based materials, i.e. the chemistry
behind the use of Ca-aluminate as biomaterials and the microstructure developed with special focus
on the nanostructure. The importance of these features for bacteriostatic and antibacterial
functioning biomaterials is treated in more details. One specific topic deals with the seemingly
problematic or contradictory but functional combination of biocompatibility and bioactivity with
antibacterial features.

MATERIALS EVALUATED
Ca-aluminate based biomaterials are based on a few of the phases in the $CaO-Al_2O_3$ system.
This system - using the cement chemistry abbreviation system – contain the following phases; C_3A,
$C_{12}A_7$, CA, CA_2 and CA_6, where C=CaO and A=Al_2O_3. Lime, CaO is an unstable alkaline
compound, whereas Al_2O_3 is one of the most stable materials whatsoever. The reactivity decreases
from C_3A to CA_6 [2]. The C_3A phase is regarded as too reactive with too high reaction
temperatures for biomaterials, and the CA_2 and especially CA_6 phases exhibit too low reaction rate.
As biomaterials the greatest interest is related to $CaO-Al_2O_3$ with molar ratio close to 1:1, i.e. the
phases CA and $C_{12}A_7$. The Ca-aluminate phase used in the testing was the mono-calcium aluminate
(CA), produced by the company Doxa AB, Sweden. After crushing, the material was jet-milled to
obtain fine-grained particles with a mean particle size below 5 μm and d(99) below 10 μm. In
addition to the main binding phases (the Ca-aluminate and water) filler particles are included to
contribute to some general properties of interest when used for different applications. The

contribution of added particles regards the microstructure (homogeneity aspects) and mechanical properties (especially hardness, Young's modulus and strength). For dental applications, additives are mainly glass particles. For orthopedic application a high density oxide is selected. For many dental applications translucency is desired, why inert particles must have a refractive index close to that of the hydrates formed [3]. A preferred high-density oxide for orthopedic application is zirconia, a material also used as a general implant material. The zirconia content will contribute to a desired radiopacity. Inert phases mixed with CA were glass particles or oxides (SiO_2 and ZrO_2) or SrF_2 depending on the applications aimed at. For low viscosity and early hardening a complementary glass ionomer system was used.

METHODS
Methods and evaluation techniques used are summarized below.

Phase and nano/microstructural studies
Studies were complemented by evaluating the chemical reactions and microstructure developed in the Ca-aluminate biomaterials. Microstructure, phase and elemental analyses were conducted using traditional SEM, TEM, HRTEM, XRD, XPS and STEM with EDX [4].

Mechanical evaluation
The Ca-aluminate materials have been evaluated concerning their mechanical properties including compressive strength, flexure strength, Young's modulus, fracture toughness and wear resistance. Detailed description of test methods is presented elsewhere [5-6].

Biocompatibility including bioactivity
The Ca-aluminate materials have been evaluated comprehensively concerning their biocompatibility and toxicological endpoints as referred in the harmonized standard ISO 10993:2003 [7], which comprises the following sections: Cytotoxicity (ISO10993-5), Sensitization (ISO10993-10), Irritation/Intracutaneous reactivity (ISO10993-10), Systemic toxicity (ISO10993-11), Sub-acute, sub-chronic and chronic toxicity (ISO10993-11), Genotoxicity (ISO10993-3), Implantation (ISO10993-6), Carcinogenicity (ISO10993-3) and Hemocompatibility (ISO10993-4).

Chemical resistance in acid environment was tested using the water jet impinging test [8].

In vitro bioactivity was evaluated using methods close to that described in ISO standard 23317 [9].

Antibacterial evaluation
The Ca-aluminate materials have been evaluated using bacterial leakage test and the direct contact test described in [10-11].

RESUTS AND DISCUSSION
The results and discussion will be presented in three parts dealing with a) the chemical reactions involved in the hydration of Ca-aluminate, b) the nanostructure developed in these reactions, and c) the consequences of this with regard to biocompatibility, bioactivity and antibacterial properties. The chemistry and the microstructure are presented in some details, since these have a great influence on the antibacterial behavior of the Ca-aluminate based biomaterials.

Chemical reactions in the Ca-aluminate system
In water environment, the Ca-aluminate cements react in an acid-base reaction to form hydrates. The reactions are temperature dependent. At body temperatures the reaction is summarized below:
- dissolution of Ca-aluminate into the liquid
- formation of ions, and

- repeated precipitation of nanocrystals (hydrates) – katoite, $CaOAl_2O_36H_2O$ (CAH_6), and gibbsite $Al(OH)_3$ (AH_3).

The main reaction for the mono Ca-aluminate phase (CA) is shown below ($H=H_2O$);

$$3CA + 12H \rightarrow C_3AH_6 + 2AH_3 \qquad \text{(Eq. 1)}$$

The reaction involves precipitation of nanocrystals on tissue walls, in the material and upon inert fillers, and repeated precipitation until the Ca-aluminate is consumed resulting in complete cavity/gap/void/ filling. This is of great importance for the antibacterial function of Ca-aluminate biomaterials.

Complementary reactions occur when the Ca-aluminate is in contact with tissue containing body liquid. Several mechanisms have been identified, which control how the Ca-aluminate material is integrated onto tissue. These mechanisms affect the integration differently depending on what type of tissue the biomaterial is in contact with, and in what state (un-hydrated or hydrated) the CA is introduced. These mechanisms are summarized as follows and described in more details elsewhere [12].

Table I: Summary of reactions mechanisms for Ca-aluminate biomaterials

Mechanism	Short Description	Comments
Mechanism 1	Main reaction, the hydration step of Ca-aluminate	Eq. 1 above
Mechanism 2	Apatite formation in presence of phosphate ions in the biomaterial	Eq. 2 below
Mechanism 3	Apatite formation in the contact zone in presence of body liquid	Eqs. 2-4 below
Mechanism 4	Transformation of hydrated Ca-aluminate (Katoite) into Apatite and Gibbsite	Eq. 5 below
Mechanism 5	Biological induced integration and in-growth	Bone formation at the contact zone.
Mechanism 6	Un-hydrated Ca-aluminate as coatings in contact with body liquid	Point welding effect, Eq. 1 above
Mechanism 7	Ca-aluminate in contact with soft tissue	Eq. 6 below

When phosphate ions or water soluble phosphate compounds are present in the biomaterial (powder or liquid) an apatite formation occurs according to the reaction

$$5Ca^{2+} + 3PO_4^{3-} + OH^- \rightarrow Ca_5(PO_4)_3OH \qquad \text{(Eq. 2)}$$

This complementary reaction to the main reaction (Mechanism 1) occurs due to the presence of Ca-ions and a basic (OH^-) environment created by the Ca-aluminate material. The solubility product of apatite is very low, pKs = 58, so apatite is easily precipitated. Body liquid contains among others the ions HPO_4^{2-} and $H_2PO_4^-$. In contact with the Ca-aluminate system and water during setting and hydration, the presence of Ca-ions and hydroxyl ions, the hydrogen phosphate ions are neutralized according to

$$HPO_4^{2-} + H_2PO_4^- + 3OH^- \rightarrow 2PO_4^{2-} + 3H_2O \qquad \text{(Eq. 3)}$$

Thereafter the apatite-formation reaction occurs

$$5Ca^{2+} + 3PO_4^{3-} + OH^- \rightarrow Ca_5(PO_4)_3OH \qquad \text{(Eq. 4)}$$

This reaction occurs upon the biomaterial surface/periphery towards tissue. The apatite is precipitated as nano-size crystals [13]. See figure 1.

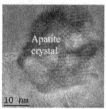

Fig. 1   Nano-size apatite formation in the the contact zone to hard tissue (bar = 10 nm)

Katoite is formed as a main phase, and is kept as katoite in the bulk material according to the mechanism 1 above. However, in long-time contact with body liquid containing phosphate ions the katoite is transformed at the interface tobody tissue into the at neutral pH even more stable apatite and gibbsite phases according to

$$Ca_3 \cdot (Al(OH)_4)_2 \cdot (OH)_4 + 2 Ca^{2+} + HPO_4^{2-} + 2 H_2PO_4^- \rightarrow$$
$$Ca_5 \cdot (PO_4)_3 \cdot (OH) + 2 Al(OH)_3 + 5 H_2O \qquad \text{(Eq. 5)}$$

The transformed contact zone is after some time constant and approximately 10 micrometer in thickness. When apatite is formed at the interface according to any of the reaction mechanisms 2-4 above, at the periphery of the bulk biomaterial, the biological integration may start. Bone ingrowth towards the apatite allows the new bone structure to come in integrated contact with the biomaterial. This is an established fact for apatite interfaces. For the CA-system the ingrowth is shown in Figure 2. The transition from tissue to the biomaterial is smooth and intricate [14].

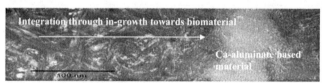

Fig. 2 Fine-tuned transition from tissue to biomaterial (bar = 500 nm)

In soft (adipose) tissue Ca-aluminate material can react and Calcite is formed [15]. The reaction is given below.

$$HCO_3^- + Ca^{2+} + OH^- \rightarrow CaCO_3 + H_2O \qquad \text{(Eq. 6)}$$

Nanostructure and nano-porosity developed

The Ca-aluminate cements exhibit an inherent property not so often considered in spite of its importance for high-strength cement materials. This deals with the huge water uptake capacity of Ca-aluminate cement. The water consumption during hydration and curing (Eq.1 above) is as high as 45 w/o water. Practically this can be utilized in development of high-strength, low-porosity materials, if appropriate w/c ratio is selected, i.e w/c ratio close to that of complete reaction of the Ca-aluminate phase. This results in microstructures with essentially no large pores, just nano-size porosity between precipitated nano-size hydrates. The total residual porosity is often as low as 5-10 %. Due to reduced porosity based on the huge water uptake ability, the Ca-aluminate material exhibits the highest strength among the chemically bonded ceramics. The inherent flexural strength is above 100 MP based on measurement of the fracture toughness, which is about 0.7 - 0.8 $MPam^{1/2}$. The actual flexural strength is controlled by external defects introduced during handling and injection of the material. In the Tables II-IV are presented typical property data of Ca-aluminate based biomaterials.

Table II. Handling property data of Ca-aluminate based biomaterials [16]

Property	Typical value
Working time, min	3
Setting time, min	5
Curing time, min	10-60*

*Can be varied within the interval depending on selection of phases,
processing agents and complementary binding phase (glass ionomer system).

Table III. Mechanical property data of Ca-aluminate based biomaterials [5-6, 17-18]

Property	Typical value
Compression strength, MPa	200
Young's modulus, GPa	10-15*
Flexural strength, MPa	40-50
Fracture toughness, $MPam^{1/2}$	0.75
Porosity after final hydration, %	5-15**

*Mainly controlled by inert fillers
** Mainly controlled by the w/c ratio of CA

Table IV. Additional property data of Ca-aluminate based biomaterials [16-17]

Property	Typical value
Thermal conductivity, W/mK	0.8
Thermal expansion, ppm/K	9.5
Corrosion resistance, water jet impinging, reduction in mm	< 0.01
Radio-opacity, mm Al	1.5-2.0*
Dimensional stability, expansion in %	0.1-0.3
Process temperature, °C	30-40

*Mainly controlled by high density inert additives

The thermal and electrical properties of Ca-aluminate based materials are close to those of hard tissue, the reason being that Ca-aluminate hydrates chemically belong to the same group as Ca-phosphates, the hard tissue of bone. Another important property related to Ca-aluminate materials is the possibility to control the dimensional change during hardening. In contrast to the shrinkage behaviour of many polymer-based biomaterials, the Ca-aluminates exhibit a small expansion, 0.1-0.3 linear-% [17]. This is important to avoid tensile stress in the contact zone between the biomaterial and tissue, and reduces significantly the risk of bacterial infiltration.

Antibacterial aspects
The nanostructure including nanoporosity developed in the Ca-aluminate biomaterial system near complete hydration conditions yields some unique properties related to how bacteriostatic and antibacterial properties develop in the biomaterial. The nano-porosity can also be used to control release of drugs incorporated in the biomaterial. The background to this is that even if the total

porosity is low, all porosity is open, thus allowing transport of molecules in the nanoporosity channels.

Based on several studies the following general reasons/conditions have been identified that describe and to some extent explain the bacteriostatic and even antibacterial features of the Ca-aluminate based biomaterials. These are summarized in Table V, and will be discussed in some details below.

Table V. Conditions contributing to antibacterial features of Ca-aluminate based biomaterials.

Condition	Description	Comments
pH	Acidic or alkaline pH interval	Antibacterial effect at pH < 6 and at pH >9
Encapsulation	Entrapping of bacteria	Bacterial growth inhibition
Surface structure	Fastening of bacteria upon the structured surface	Bacterial growth inhibition
F-ion presence	F-ions act as OH-ions	Antibacterial effect even at neutral conditions

The surprising finding in studies recently performed [19] show that the bacteriostatic and antibacterial properties of the Ca-aluminate biomaterial may not just be related to pH, but also to the hydration procedure and the microstructure obtained. This also to some extent is an answer why highly biocompatible and even bioactive biomaterials can combine apparently contradictory features such as biocompatibility, bioactivity and apatite formation and environmental friendliness with bacteriostatic and antibacterial properties. The identified areas related to antibacterial features are discussed below.

pH

Bacteria have great problems to survive at low pH, < 6, and at high pH, > 9. Antibacterial features have been studied for the chemically bonded bioceramics Ca-aluminate based biomaterials and Ca-silicate based biomaterials [19-20]. For pure Ca-aluminate based materials the pH during the initial hardening is high, approximately 10.5. The antibacterial property is obvious. However, even for Ca-aluminate biomaterial with the glass ionomer system, where the cross-linking poly acrylic acid yields the system an initial low pH, approximately pH 5 and < 7 for 1 h, and then during final curing and hardening a pH of 8-9, antibacterial features appear. So it quite clear that pH-conditions are not the only reasons for the antibacterial situation in Ca-aluminate based biomaterials.

Encapsulation

The main reaction for the mono Ca-aluminate phase (CA) shown above (Eq. 1) involves precipitation of nanocrystals on tissue walls, in the material and upon inert fillers, and repeated precipitation until the Ca-aluminate is consumed resulting in complete cavity/gap/void/ filling. This reaction will guarantee that the nanostructure will be free of large pores, meaning no escape of bacteria within the original liquid, paste or dental void, during the hydration. The nanocrystals will participate on all walls, within the liquid, and on all inert particles and on bacteria within the original volume. The formation of nanocrystals will continue to all the void is filled. The bacteria will be totally encapsulated and will be chemically inactivated. In Fig. 2 above and in Fig. 3 below are shown how well the nanosize hydrates are attached to both biological material and other biomaterials. In Fig. 3 the nanostructural integration of a Ca-aluminate based material is shown - in this case the contact area to a titanium based implant [21]. The contact zone between the biomaterial and tissue and other implant materials is kept intact as there is no shrinkage in the formation of the zone - just a slight expansion. Thus no tensile stresses develop. The reason for later opening in contact zones and bacterial invasion reported in the dental literature for polymer composites is ascribed the shrinkage of these biomaterials.

Fig. 3 The nanostructure integration at the contact zone between a titanium implant (top) and a Ca-aluminate hydrated paste (bottom), High-resolution TEM (bar = 10 nm)

Surface structure

The bacteriostatic and antibacterial properties are in addition to pH-conditions and the nanostructural entrapping mechanism also related to the surface structure developed of the hydrated biomaterial. The nanoparticle/crystal size of hydrates are in the interval 15-40 nm with a nanoporosity size of 1-4 nm. The number of pores per square micrometer is at least 500, preferably > 1000 [19]. The number of nanopores will thus be extremely high, which will affect the possibility of catching and fasten bacteria to the hydrate surface – an analogue to how certain peptides may function as antibacterial material due to a structure with nano-size holes within the structure. This may also provide a long-term antibacterial activity after the initial hydration.

OH and F ions

The size of F-ions is almost the same as that of OH-ions, approximately 1.4 Å. It is proposed that the antibacterial effect may partly be related to the presence of F-ions. These may have the same effect upon the bacteria as a high pH, i.e. a high hydroxyl concentration. The F-containing slowly resorbable glass and the Sr-fluoride may thus contribute to the antibacterial features.

CONCLUSION AND OUTLOOK

The oral environment is a body area with high bacterial activity, which causes the most common dental problem - caries formation. The main hydration reactions in the Ca-aluminate system, the general stability of the hydrates formed, and the nanostructure developed make these materials suitable as injectable biomaterials into tissue, within odontology as cements and restoratives, and within orthopedics as augmentation materials [22-23]. The nanostructure of the materials also makes these biomaterials potential as drug delivery carriers [24].

The antibacterial properties of the Ca-aluminate system during hydration and as a permanent implant material are due to the following;
- pH during initial hardening – antibacterial effect at low and high pH
- F-ions presence – similarity in size to hydroxyl ions
- The nanostructure developed – entrapping of bacteria and growth inhibition
- The surface structure of hydrated Ca-aluminate – fastening of bacteria to the surface and growth inhibition.

The antibacterial features of the Ca-aluminate based materials do not affect the established biocompatibility and bioactivity of these biomaterials.

ACKNOWLEDGEMENT
Results presented in this paper are based on two decades of research within Doxa AB, and the Doxa personnel are greatly acknowledged.

REFERENCES

[1] A. Ravaglioli and A. Krajewski, *Bioceramics,* Ed Chapman and Hall, 1992

[2] R. J. Mangabhai, Calcium Aluminate Cements, *Conf. proceeding,* Chapman and Hall, 1990

[3] H. Engqvist, J. Loof, S. Uppstrom, M. W. Phaneuf, J. C. Jonsson, L. Hermansson and N-O. Ahnfelt, Transmittance of a bioceramic Calcium aluminate based dental restorative material, *Journal of Biomedical Materials Research Part B: Applied Biomaterials* vol. 69 no. 1 (2004), 94-98

[4] H. Engqvist, J-E. Schultz-Walz, J. Loof, G. A. Botton, D. Mayer, M. W. Phaneuf, N-O. Ahnfelt, and L. Hermansson, Chemical and biological integration of a mouldable bioactive ceramic material capable of forming apatite in vivo in teeth, *Biomaterials* Vol 25 (2004), 2781-2787

[5] J. Loof, H. Engqvist, L. Hermansson, and N. O. Ahnfelt, Mechanical testing of chemically bonded bioactive ceramic materials, *Key Engineering Materials* Vols. 254-256 (2004), 51-54

[6] L. Hermansson, L Kraft, K Lindqvist, N-O Ahnfelt, H Engqvist, Flexural Strength Measurement of Ceramic Dental Restorative Materials, *Key Engineering Materials*, Vols. 361-363 (2008), 873-876

[7] ISO standard 10993:2003

[8] EN 29917:1994/ISO 9917:1991

[9] ISO Standard 23317 (2007)

[10] C. Pameijer, O. Zmener, S. A. Serrano and F. Garcia-Godoy, Sealing properties of a calcium aluminate dental cement, *American Journal of Dentistry,* Vol 23 121-124 (2010)

[11] E. Weiss, M. Shalhav and Z. Fuss, Assessement of antibacterial activity of endodontic sealers by direct contact test, *Endod. Dental Traumatol.* Vol 12 179-84 (1996)

[12] L. Hermansson, J. Lööf, T. Jarmar, Integration mechanisms towards hard tissue of the Ca-aluminate based biomaaterials, *Key Engineering Materials.* 2009;396-398:183-186

[13] L. Hermansson, H. Engqvist, J. Lööf, G. Gómez-Ortega and K. Björklund, Nano-size biomaterials based on Ca-aluminate, Adv. In Sci. and Techn., *Key Eng. Mater.* Vol. 49, 21-26 (2006)

[14] L. Hermansson, J. Lööf, and T. Jarmar, Injectable ceramics as biomaterials – today and tomorrow, *Proc ICC2* Verona, (2008)

[15] N. Axén, L.-M. Bjursten, H. Engqvist, N.-O. Ahnfelt and L. Hermansson, Zone formation at the interface between Ca-aluminate cement and bone tissue environment, *Ceramics, Cells and Tissues, 9th Annual Seminar & Meeting,* Faenza (2004)

[16] L. Hermansson, A. Faris, G. Gómez-Ortega, E. Abrahamsson and J. Lööf, Calcium Aluminate based dental luting cement with improved properties – an overview, *Ceramic Eng. And Sci. Proc.* Vol 31 27-38 (2010)

[17] L. Kraft, Calcium aluminate based cement as dental restorative materials, *Ph D Thesis,* Faculty of Science and technology, Uppsala University, Sweden. 2002

[18] J. Lööf, Calcium-aluminate as biomaterial: Synthesis, design and evaluation, *Ph D Thesis,* Faculty of Science and Technology, Uppsala, University, Sweden. 2008

[19] Patent Pending, Doxa AB (2010)

[20] Patent RU 2197940 (2003)

[21] L.Hermansson, Nanostructural Chemically Bonded Ca-Aluminate Based Bioceramics, Chapter 3 in *Biomaterials,* Publ by INTECH (2011)

[22] J. Lööf, A. Faris, L. Hermansson, and H. Engqvist, In Vitro Biomechanical Testing of Two Injectable Materials for Vertebroplasty in Different Synthetic Bone, *Key Engineering Materials* Vols. 361-363, (2008), 369-372

[23] T. Jarmar, T. Uhlin, U. Höglund, P. Thomsen, L. Hermansson and H. Engqvist, Injectable bone cements for vertebroplasty studied in sheep vertebrae with electron microscopy, *Key Engineering Materials*, Vols. 361-363 (2008), 373-37620)

[24] L. Hermansson, Chemically bonded bioceramic carrier systems for drug delivery. Publ. in *Advanced Ceramics and Composites, ICACC,* Daytona Beach (2010)

POTENTIAL TOXICITY OF BIOACTIVE BORATE GLASSES IN-VITRO AND IN-VIVO

Steven B. Jung[1], Delbert E. Day[2], Roger F. Brown[3], and Linda F. Bonewald[4]

[1]MO-SCI Corporation, Rolla, MO, 65401
[2]Graduate Center for Materials Research, Materials Science and Engineering Department, Missouri University of Science and Technology, Rolla, MO, 65409-1170
[3]Department of Biological Sciences, Missouri University of Science and Technology, Rolla, MO, 65409-1170
[4]Department of Oral Biology, School of Dentistry, University of Missouri Kansas City, Kansas City, MO 64108-2784

ABSTRACT

   Potential toxicity of a bioactive borate glass was evaluated using *in-vivo* animal models in soft tissue and bone and *in-vitro* cell culture using MLOA5 late osteoblast/early osteocytes cells. No toxicity was found between bioactive borate glass and subcutaneous tissue, in the liver, and only normal incidental changes in rat kidney. Bone growth across porous scaffolds composed of randomly oriented borate glass fibers was significantly higher than for a scaffold composed of a borosilicate or a silicate bioactive glass (13-93) fibers after 12 weeks *in-vivo*, $p<0.05$. *In-vitro* cell culture on bioactive glasses showed that under static culture conditions, borate glass disks tended to inhibit the growth of MLOA5 cells. The present work, along with literature data, show that bioactive borate glasses are biocompatible *in-vivo* and are not toxic to adjacent hard or soft tissues, or internal organs such as the kidney and liver, at the relatively high, estimated concentrations ($<\sim126mg/kg/day$). Based on the present results and literature data, ions released from the borate glasses such as alkali or boron were not toxic in a dynamic environment such as the body and should be considered for use in humans and other mammals for soft and hard tissue engineering applications.

INTRODUCTION

   In the world of biomaterials, the lifecycle of a new material starts with typical material design (composition, microstructure, surface texture, etc) followed by a multitude of in-vitro experiments to determine bioactivity (hydroxyapatite (HA) formation [1,2], cell adhesion and proliferation [3,4], or toxicity [5]. Depending on the results of these survey experiments, the material either passes to go on to the next battery or experiments, which may be additional in-vitro characterization or in-vivo implantation, or it fails, and its back to the drawing board to start over. This sort of evaluation is straight forward, relatively easy to evaluate, and for the most part been effective for decades [6]. This evaluation method requires some hard and fast assumptions such as: in-vitro analysis directly correlates to in-vivo performance, in-vitro conditions are a comparable surrogate for the in-vivo environment, and in-vitro toxicity disqualifies a material from further analysis.

   Materials that will pass the in-vitro conditions tend to be inert or slow reacting. Case in point, cell cultures are done primarily in inert plastic dishes. The cells grow well across the surface, and the dish does nothing. In contrast, the materials of the future are often described as bioactive [2], resorbable, degradable [7], biocompatible [8], antimicrobial [8], etc . These words typically mean the material will interact with the in-vivo environment and either form HA [5,9], or they release ions to signal cellular interactions, or perhaps they just degrade in the body fluids. Regardless of the function, these new materials are different than the inert plastic cell culture dish, and arguably should be treated differently when evaluating. The main focus of the present work is to show how the traditional in-vitro evaluation methods of bioactive glasses overlooked a family of bioactive borate glasses that in-vivo has proven effective as new treatment options for treating bone defects and chronic soft tissue wounds such as diabetic ulcers [10].

EXPERIMENTAL

Glass Wafer and Scaffold Preparation
    Bioactive glasses were batched from reagent grade chemicals, melted in platinum crucibles between 1050°C and 1400°C for 2 hours, and cast into solid rods (10mm diameter) for making wafers, or glass fibers were pulled from the melt to make three dimensional porous fiber scaffolds. The scaffolds use for in-vitro analysis consisted of 70mg of glass fibers for (100 to 300μm diameter, 1 to 3mm in length) that were placed in a mullite mold and heat treated for 45 minutes at 575°C to 700°C depending on composition to produce a 7mm diameter by 2mm thick scaffold with ~50% open porosity. An example of a 13-93B3 sintered fiber scaffold is shown in Fig 1. Similar scaffolds were used for in-vivo analysis, but they were 4mm in diameter and 1.5mm thick. A complete list of glass compositions and scaffold processing parameters are shown below in Table I.

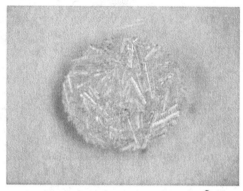

3mm

Figure 1 - Optical micrograph of an as-made 13-93B3 scaffold after heating for 45 minutes at 575°C. The scaffold has a nominal diameter and thickness of 7 ± 0.1mm and 2 ± 0.2mm, respectively, with an average open and interconnected porosity of 51 ± 2%, n=12 [11].

Table I. Target Glass Compositions (wt%) and Scaffold Processing Parameters

Glass ID	SiO$_2$	Na$_2$O	MgO	K$_2$O	CaO	P$_2$O$_5$	B$_2$O$_3$	Melt Temp °C	Sintering Temp °C	Sintering Time (min)
45S5	45	24.5	0	0	24.5	6	0	1350	NA	NA
13-93	53	6	5	12	20	4	0	1400	700	45
13-93B1	35.3	6	5	12	20	4	17.7	1300	625	45
13-93B3	0	6	5	12	20	4	53	1050	575	45

In-Vitro Cell Culture
    All wafers or scaffolds were washed twice with ethyl alcohol and heat sterilized at 250°C for a minimum of three hours in a small oven prior to cell seeding. MLOA5 late osteoblast/early osteocyte cells were seeded on the wafers or scaffolds for cell proliferation measurements by MTT analysis at two, four, and six days. Fifty thousand MLOA5 cells were seeded on each wafer/scaffold and cultured

in Sigma α-MEM 0644 media (Sigma Aldrich, St. Louis, MO) at 37°C and in 95% air/5% CO_2 atmosphere. Wafers/scaffolds cultured two, four, and six days were incubated the last four hours in 400µl of serum-free medium containing 100µg tetrazolium salt MTT to permit visualization of metabolically active cells on the scaffolds. After incubation, the scaffolds were rinsed with sterile PBS and dried.

In-Vivo Scaffold Implantation in Rat Calvaria

X-ray amorphous fiber scaffolds of 13-93, 13-93B1, and 13-93B3 glasses (see Table 1) with an open porosity of 50±2% and similar pore structure, and 100 to 200µm diameter loose particles of 45S5 glass were implanted in 4mm diameter calvaria defects of rats for 12 weeks. The rats were anesthetized with intravenous drugs prior to the surgery. Two 4mm diameter defects were made with a high speed dental bur and placed on each side of the sinus as shown in the schematic in Fig 2. The fiber scaffolds or loose particles of 45S5 were implanted in the defects, and the skin was closed. The calvaria implantation and micro-CT measurements were done at the University of Missouri – Kansas City School of Dentistry. Histomorphometry analysis of the 12 week scaffolds and loose particles for quantization of total bone growth was compared by one-way analysis of variance (ANOVA).

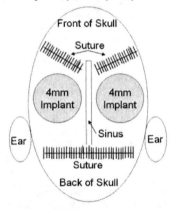

Figure 2 - Schematic showing the scaffold placement in the calvaria of a rat. The scaffolds implanted measured 4mm in diameter and 1.5mm thick as shown in Fig 2. Each scaffold was placed between the surrounding sutures and the sinus was avoided to reduce surgical complications.

RESULTS AND DISCUSSION

In-vitro Analysis of Bioactive Glass

Figure 3 shows the cell morphology typical of a static in-vitro cell cuture of MLOA5 cells on wafers of 13-93, 13-93B1, and 13-93B3 bioactive glasses after one and three days. The silicate 13-93 reacts at a rate of 2 to 3µm a week in-vivo, so this glass is relatively slow reacting, and the cells appear well adhered and spread across the surface of the wafer at both one and three days. The borosilicate 13-93B1 reacts at 3 to 4 µm per week, so similar to 13-93, and the morphology of the cells appears flat and spread across the wafer surface and the number of cells has multiplied indicating the environment is suitable for cell viability and proliferation. The 13-93B3 glass however did not support cell growth as indicated by the relatively few balled up, and likely non-viable cells shown at days one and three. The in-vivo reaction rate of the borate 13-93B3 is 25 to 30 µm a week, so an order of magnitude higher than the other two glasses. By this test alone, the 13-93B3 glass appears toxic and likely a poor choice for a biomaterial.

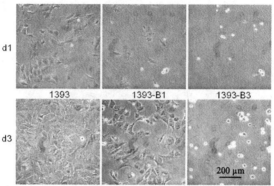

Figure 3 – Static in-vitro cell culture on bioactive glass wafers after one and three days.

The next analysis shown in Fig 4 is a static in-vitro culture of MLOA5 cells on three dimensional porous fiber scaffolds, similar to the scaffold shown in Fig 1, composed of 13-93 (top) and 13-93B3 glass (bottom) and stained with MTT. The MTT is purple, so the darkening of the 13-93 scaffold at day 4 and day 6 indicates there was an increase in viable cells as a function of time, and the results are comparable with the wafer test from Fig 3. The 13-93B3 scaffold at day 6 did darken from day 2 indicating that the cells were multiplying and viable, but overall performed poorly when compared to 13-93. As a point of reference, the pH increased from the original 7.4 to ~8 in the 13-93 group and ~9 in the 13-93B3 group. This analysis again shows that 13-93B3 is likely not a good choice for a scaffold material.

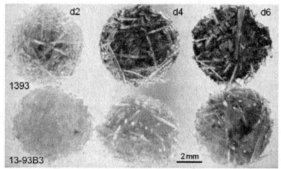

Figure 4 – Static in-vitro cell culture on bioactive glass fiber scaffolds (13-93 top and 13-93B3 bottom) after two, four, and six days. The scaffolds were stained with MTT to show viable cells (purple color).

In-vivo Analysis in a Rat Calvaria Model

After the poor in-vitro results, most researchers would stop testing the 13-93B3 glass as it was showing toxicity to cells and there was another better choice in 13-93 or 13-93B1 to continue with the scaffold research. The most important parameter to biomaterials however is in-vivo performance, so instead of eliminating 13-93B3, it was tested against 13-93, 13-93B1, and 45S5 in a rat calvaria model. Due to the difficulties with sintering 45S5, it was used in particulate form.

The present study is the first to compare similar scaffold microstructures composed of borate and silicate glasses. On the average, the scaffold with the most bone regenerated across the scaffold top and bottom surfaces after 12 weeks as measured by micro-CT (MicroCT-40, Scanco Medical AG) was the 13-93B3, and 13-93B3 the only scaffold type to have bone completely cover the bottom side of a scaffold. The 13-93B3 scaffold had significantly more bone than the 13-93 or 13-93B1 scaffolds by histomorphometry analysis (1 way ANOVA).

The 45S5 glass was implanted in particulate form as previously mentioned because the glass is difficult to heat treat and form a porous scaffold without crystallization. The objectives of the calvaria experiment were to determine the effect of glass composition on bone growth and to determine if a scaffold with interconnected pores was better for bone growth than loose particles. The facts of the experiment are that the borate glass had significantly more bone ($p<0.05$) than the borosilicate or silicate glass scaffolds and the borate scaffold had on the average more bone than the 45S5 loose particles (Fig 6).

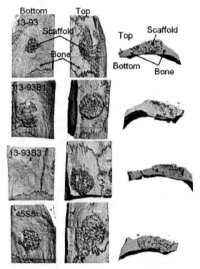

Figure 5 - Micro-CT images of the top and bottom of porous randomly oriented fiber scaffolds composed of bioactive glass fibers (13-93, 13-93B1, 13-93B3) and 45S5 particulates (100 to 200μm) after 12 weeks in 4mm rat calvaria defects. The scaffold and bone are labeled for the 13-93 scaffold for the top, bottom, and cross sectioned views. Note that the 13-93B3 scaffold is not visible in the bottom image (left). Bone had completely grown across the bottom of the 13-93B3 scaffold [11].

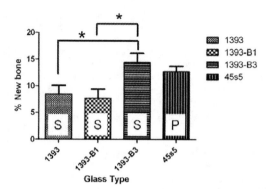

Figure 6 – Total bone growth into porous bioactive glass scaffolds (S), (45S5 was in particulate (P) form (100 to 200μm)) after 12 weeks in rat calvaria (n=4) as determined by histomorphometry analysis [11]. Statistical significance represented by *($p<0.05$) [11].

According to the in-vivo histomorphometry[11], the exact opposite result occurred with respect to bone growth as would be expected based on the in-vitro results and toxicity. The schematic in Fig 7 illustrates what is occurring at the cellular level at the glass to cell interface that may cause this deviation. On the left, the static in-vitro model shows the glass reacts with the surrounding media and alkali and boron are released from the glass, but due to the lack of circulation, the majority of the ions stay close to the glass surface which changes the local media composition and pH. This is probably why the 13-93B3 glass does so poorly in the static in-vitro analysis. On the right of Fig 7, the dynamic condition is modeled. As the glass reacts with the surrounding fluids, the alkali and boron are released the same as in the in-vitro condition, but the fluids at the glass surface are constantly being replenished, so the production of a local toxic environment is significantly reduced by removal of the ions. This constant replenishment is likely why the 13-93B3 glass performed better in-vivo. Based on the in-vivo reaction rates of the 13-93 , 13-93B1, and 13-93B3 glasses discussed earlier compared with an in-vivo reaction rate for 45S5 of 5 to 6μm/wk[11], and the known effect calcium has on stimulating bone cell migration [12], the result of more bone present in the faster reacting borate glass scaffold than the silicate and borosilicate glasses seems a more likely result.

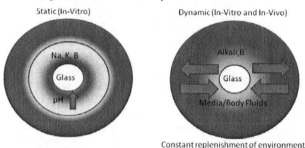

Figure 7 – Model showing the environmental conditions present at the surface of a bioactive glass under static (left) and dynamic conditions (right).

In-vivo Histology on Kidney and Liver

Although the 13-93B3 glass was shown to grow bone in the calvaria defect, the ions released from the glass, primarily boron, may have caused systemic tissue damage to other organs that process and remove them. The results shown in Table II show that animals implanted with 16 scaffolds (~126mg/kg/day H_3BO_3) in subcutaneous tissue (70mg each) showed no significant histological difference from the control animals (no implanted scaffolds) in terms of kidney or liver damage [11]. This result further improves the chances of the 13-93B3 glass as a useful material for tissue engineering.

Table II. Histological Assessment of Kidney and Liver from Rats Implanted with 13-93B3 Scaffolds

Scaffolds/Animal	16	16	16	16	12	12	12	12	8	8
Tissue Finding										
Kidney	+	+	+	-	+	+	-	+	+	+
Degeneration, Tubular	1M	1M	-	-	-	-	-	-	1M	-
Protein Casts, Tubular	1M	1M	1M	-	-	-	-	-	1M	1M
Nephocalcinosis	1M	-	-	-	1F	1M	-	1M	1F	-
Liver	-	-	-	-	-	-	-	-	-	-

Scaffolds/Animal	8	8	4	4	4	4	Control	Control	Control	Control
Tissue Finding										
Kidney	-	+	+	+	+	+	-	+	+	+
Degeneration, Tubular	-	-	-	-	1M	1M	-	-	-	-
Protein Casts, Tubular	-	1M	1M	1M	1M	1M	-	1M	1M	1M
Nephocalcinosis	-	-	-	-	-	-	-	-	-	-
Liver	-	-	-	-	-	-	-	-	-	-

Key: - = negative/no significant finding, + = positive finding present, 1 = minimal severity, 2 = mild severity, 3 = moderate severity, 4 = marked severity, F = focal, M = multifocal, D = diffuse

Dynamic In-vitro Analysis of Bioactive Borate Glass Scaffolds

To test the hypothesis that a dynamic environment will improve the in-vitro cell culture of MLOA5 cells on a porous 13-93B3 scaffold, three scaffolds were placed in a dynamic environment while three were cultured statically. The dynamic environment was established by a platform rocker that tilted back and forth to agitate the culture media once every five minutes. The scaffolds were stained with MTT after two days in culture, and the scaffolds are shown below in Fig 8. The static scaffolds looked similar to the two day scaffold from Fig 4, and showed the presence of few viable cells. The dynamic environment is visibly darker in color than the static scaffolds indicating a higher number of viable cells were present. This test while relatively simple shows that the in-vitro environment can be modified by simple agitation to more closely mimic the in-vivo environment. Highly reactive materials, such as the 13-93B3 glass, need to be tested in a way that will fairly and more accurately represent what might be expected in-vivo.

Figure 8- MLOA5 cell seeded 13-93 B3 fiber scaffolds labeled with MTT after 2 day incubation; static (bottom) and dynamic (top) conditions.

Treatment of a Human Chronic Soft Tissue Wound with Bioactive Borate Glass

Based on the promising in-vivo analysis of the 13-93B3 glass scaffolds in the rat calvaria, a related borate glass composition was made into a nano-fiber material with a consistency similar to a cotton ball as shown in Fig 9 [10]. This nano-fiber material (200nm to ~2μm diameter) was specially designed to mimic the microstructure of the initial stage of healing (a fibrin clot) while releasing ions such as copper and zinc to aid in the healing process. The fibers were made by MO-SCI Corporation by a proprietary glass fiber fabrication method.

Figure 9 – Bioactive borate glass doped with copper and zinc and made into a nano-fiber material for wound healing applications. Notice the blue color from the addition of copper to the glass.

With hospital IRB (internal review board) approval, the borate glass fibers were used to treat a chronic wound located on the lower back of a paraplegic patient after treatment with all other commercial bandages and dressing failed to heal the wound. Typical treatment of the wound was an application of the fiber two to three times per week with minimal debridement until full closure [10].

The fiber was covered with a traditional bandage to hold the fiber in place and absorb excess moisture from the wound.

Images from before and after treatment with the fiber are shown below in Fig 10. On the left of Fig 10, the location of the wound is pictorially represented. The initial wound dimensions were 15cm x 12cm x 1cm deep with areas of undermining (empty cavity below surface of skin) were present on the initial day of treatment. After approximately five months of treatment began, the wound had filled with new tissue and healed with minimal amounts of scarring.

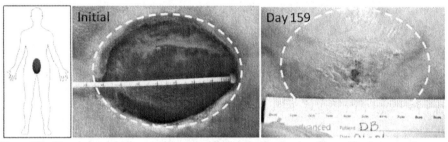

Figure 10 – Chronic lower back wound treated with bioactive borate glass fibers. The image on the left shows the approximate location on the body, the center image shows the wound prior to treatment, and the image on the right shows the wound after 159 days of treatment. The white dashed ovals were added to guide the eye.

CONCLUSION

The present work, while illustrated with bioactive borate glasses, showed how in-vitro analysis has the potential to generate data which can easily misrepresent the potential properties of a biomaterial. While in-vitro analysis is important, it must be used cautiously and the experimenter should understand what the results really mean and what might be the underlying cause for any outcome, positive or negative. The present work makes a case for new and improved test methods for determining the toxicity and effectiveness of reactive materials in-vitro to better mimic what might be expected in-vivo. In the case of the bioactive borate glasses, in-vitro analysis indicated a toxic material that most people would have abandoned, but the in-vivo data is telling a completely different story with potential to help millions of people with hard and soft tissue wounds and deficiencies.

ACKNOWLEDGEMENT

The authors would like to thank the US Army Medical Research and Material Command (Grant No. W81XWH-08-1-0765) for financial support and the Phelps County Regional Medical Center IRB for consideration and allowance of the borate glass fiber treatment.

REFERENCES

1. Jung SB, Day DE. Conversion kinetics of silicate, borosilicate, and borate bioactive glasses to hydroxyapatite. Physics and Chemistry of Glasses 2009;50(2):85-88.
2. Hench LL, Hench JW, Greenspan DC. Bioglass: A short history and bibliography. Journal of Australian Ceramic Society 2004;40:1-42.
3. Rahaman MN, Mao JJ. Stem Cell-Based Composite Tissue Constructs for Regenerative Medicine. Biotechnology and Bioengineering 2005;91(3):261-284.

4.   Fu Q, Rahaman MN, Bal BS, Brown RF, Day DE. Mechanical and InVitro Performance of 13-93 Bioactive Glass Scaffolds Prepared By a Polymer Foam Replication Technique. Acta Biomaterialia 2008;4(6):257-293.
5.   Brown RF, Rahaman MN, Dwilewicz AB, et al. Effect of borate glass composition on its conversion to hydroxyapatite and on the proliferation of MC3T3-E1 cells. Journal of Biomedical Materials Research 2008;88A:392-400.
6.   Hench LL. The story of Bioglass. Journal of Materials Science: Materials in Medicine 2006;17(11):967-978.
7.   Merolli A, Tranquiilli P, Guidi PL, Gabbi C. Comparison in In Vivo Response Between a Bioactive Glass and a Non-Bioactive Glass. Journal of Materials Science: Materials in Medicine 2000;11:219-222.
8.   Clupper DC, Hench LL. Bioactive Response of Ag-doped Tape Cast Bioglass 45S5 Following Heat Treatment. Journal of Materials Science: Materials in Medicine 2001;12:917-921.
9.   Huang W, Rahaman MN, Day DE, Li Y. Mechanisms for Converting Bioactive Silicate, Borate, and Borosilicate Glasses to Hydroxyapatite in Dilute Phosphate Solution. Physics and Chemistry of Glasses 2006;47B(6):1-12.
10.   Jung SB. Wound Healing Power of Glass. Nanotech Insights 2011;2(3):2-4.
11.   Jung SB. Borate Based Bioactive Glass Scaffolds for Hard and Soft Tissue Engineering. Rolla, MO: Missouri University of Science and Technology; 2010.
12.   Yamaguchi T, Chattopadhyay N, Kifor O, Robert R. Butters J, Sugimoto T, Brown EM. Mouse Osteoblastic Cell Line (MC3T3-E1) Expresses Extracellular Calcium (Ca2+) Sensing Receptor and Its Agonists Stimulate Chemotaxis and Proliferation of MC3T3-E1 Cells. Journal of Bone and Mineral Research 1998;13(10):1530-1538.

FABRICATION OF CARBONATE APATITE-PLGA HYBRID FOAM BONE SUBSTITUTE

Girlie M. Munar[1], Melvin L. Munar[1], Kanji Tsuru[1], Shigeki Matsuya[2] and Kunio Ishikawa[1]
[1] Department of Biomaterials, Faculty of Dental Science, Kyushu University, Fukuoka, Japan
[2] Sections of Bioengineering, Fukuoka Dental College, Fukuoka, Japan

ABSTRACT

Porous carbonate apatite (CO_3Ap) foam is a potential bone substitute material since it approximates the morphology and mineral phase of bone. One drawback however, is the poor mechanical properties for sufficient handling. A useful method to improve the mechanical property is by reinforcing it with biodegradable polymer. This study reports the preparation of carbonate apatite-PLGA (CO_3Ap-PLGA) hybrid foam with improved mechanical strength and osteoconductivity. CO_3Ap foam was prepared by hydrothermal treatment of α-tricalcium phosphate (αTCP) foam in carbonate solution at 150°C for 24 hours. CO_3Ap powder was synthesized from vaterite ($CaCO_3$) and disodium hydrogen phosphate (Na_2HPO_4) aqueous solution at 37°C. The obtained CO_3Ap powder was mixed with 10wt% PLGA solution then reinforced on CO_3Ap foam using freeze-vacuum technique. The obtained CO_3Ap-PLGA hybrid foam showed interconnecting porous structure with average porosity of 85%. Compressive strength of CO_3Ap-PLGA hybrid foam was as high as 0.35 MPa when compared to that of CO_3Ap foam at 0.01 MPa. X-ray diffraction and FT-IR showed CO_3Ap as the primary mineral phase. In conclusion, CO_3Ap-PLGA hybrid foam with improved mechanical properties and approximates the mineral composition and morphology of the cancellous bone can be a potential bone substitute or scaffold for tissue engineering.

INTRODUCTION

Calcium phosphate foam has gained much attention due to their usefulness as bone substitute and scaffold for tissue engineering[1,2]. Among them, CO_3Ap foam is a good candidate since it has been demonstrated that the biological apatite of bone, dentin and tooth enamel contains carbonate[3,4]. In addition, osteoclast cells can resorb CO_3Ap[5,6]. CO_3Ap is therefore an ideal bone substitute since it could be remodeled and replaced by new bone.

However, CO_3Ap foam, like other related calcium phosphate foam suffers from low mechanical strength required for basic handling and manipulation such as shaping and cutting to conform to the bone defect size and shape. Through mimicking the bone structure, a composite approach is a useful way to improve the mechanical properties of such biomaterial. After all, the bone is a composite of organic and inorganic phases, which provides it with strength, rigidity and partial elasticity.

The goal of this study was to fabricate a bioresorbable porous hybrid foam from organic polymer and inorganic CO_3Ap. The hybrid foam was prepared by reinforcing a CO_3Ap matrix with a composite solution made up of low-crystalline CO_3Ap powder and copolymer of poly(DL-lactide-co-glycolide) (PLGA). Scanning electron microscopy, x-ray diffraction, fourier transform infrared spectroscopy, compressive strength and porosity were used to characterize the CO_3Ap-PLGA hybrid foam. Pure CO_3Ap foam was also characterized for comparison.

MATERIALS AND METHODS

Preparation of CO_3Ap foam

CO_3Ap foam was prepared based on compositional transformation of α–tricalcium phosphate (αTCP) foam precursor[7,8]. Briefly, polyurethane foam templates were coated with αTCP slurry, dried overnight at 60°C and sintered at 1,500°C for 5 hours. The obtained αTCP foam was immersed in

4mol/L ammonium carbonate $(NH_4)_2CO_3)$ solution and hydrothermally-treated at 150°C for 24 hours for compositional transformation to CO_3Ap foam.

Preparation of mixed CO_3Ap-PLGA solution
Low-crystalline CO_3Ap powder was synthesized from vaterite $(CaCO_3)$ and disodium hydrogen phosphate (Na_2HPO_4) aqueous solution at 37°C for 1 week. The obtained CO_3Ap powder was mixed with 10wt% PLGA solution so that the concentration of CO_3Ap powder particles in the solution was 66.7wt%.

Reinforcement procedure
5 mL of the CO_3Ap-PLGA solution was transferred into a glass tube and immediately frozen in liquid nitrogen. The CO_3Ap foam was placed on top of the frozen CO_3Ap-PLGA solution and immediately placed under vacuum. Once the frozen CO_3Ap-PLGA solution had liquefied, the vacuum was released and the specimen was held under normal atmospheric pressure. After 30 minutes, the reinforced CO_3Ap foam was removed from the CO_3Ap-PLGA solution, blotted with highly absorbent paper and blown with compressed air to remove excess solution. Finally, the specimens were dried at 37°C for 1 week.

Characterization
Morphology and surface structure were observed using a scanning electron microscope (SEM). Composition was analyzed using x-ray diffraction (XRD) and fourier transform infrared spectroscopy (FT-IR). Compressive strength was measured using a universal testing machine. Porosity was calculated using the bulk density of the foam and theoretical density of the strut.

RESULTS
Structure of the polyurethane foam, CO_3Ap foam and CO_3Ap-PLGA hybrid foam were basically the same. In other words, they demonstrated fully interconnected porous structure similar to cancellous bone. At higher magnification, surface of CO_3Ap-PLGA hybrid foam was mostly CO_3Ap particles. Surface of CO_3Ap foam on the otherhand, was typically made up of large size polygonal-shaped crystals.

Cross-section of the CO_3Ap-PLGA hybrid foam revealed that the thickness of the reinforced CO_3Ap-PLGA solution was 50 μm.

XRD patterns of CO_3Ap-PLGA hybrid foam and CO_3Ap foam showed typical apatitic pattern. However, peaks were relatively broader in the case of CO_3Ap-PLGA hybrid foam. This may be caused by the use of low crystalline CO_3Ap for the reinforcement of CO_3Ap-PLGA layer whereas CO_3Ap foam showed relatively high crystallinity. No additional peaks were shown other than those assigned for apatite for both specimens.

FT-IR spectra of CO_3Ap-PLGA hybrid foam and CO_3Ap foam were typical for CO_3Ap. In other words, peaks assigned to carbonate were detected at around 870, 1410 and 1460 cm^{-1} on CO_3Ap-PLGA hybrid foam and CO_3Ap foam. The absorption bands at 1410 and 1460 cm^{-1} also indicated that B-type CO_3Ap was obtained in this study. In the case of FT-IR spectra for CO_3Ap-PLGA hybrid foam, peaks assigned to PLGA were also observed at around 1770 and 750 cm^{-1}.

Table 1 shows the compressive strength and porosity of CO_3Ap-PLGA hybrid foam and CO_3Ap foam. Compressive strength of CO_3Ap-PLGA hybrid foam (0.4 MPa) was higher than CO_3Ap foam. In contrast, porosity of CO_3Ap-PLGA hybrid foam was lower than CO_3Ap foam.

Table 1. Compressive strength and porosity of CO_3Ap-PLGA hybrid foam and CO_3Ap foam.

Specimen	Compressive Strength (MPa)	Porosity (%)
CO_3Ap-PLGA hybrid foam	0.4 ± 0.1	85 ± 1
CO_3Ap foam	0.01 ± 0.00	95 ± 1

DISCUSSION

In this study, CO_3Ap-PLGA hybrid foam aimed as bone substitute material was fabricated. To prepare the CO_3Ap-PLGA hybrid foam, the CO_3Ap foam was initially used as the matrix. Then a composite of poly(DL-lactide-co-glycolide) (PLGA) and low-crystalline CO_3Ap particles were reinforced on the CO_3Ap foam matrix. The obtained CO_3Ap-PLGA hybrid foam had better mechanical properties than CO_3Ap foam. As shown in Table 1, compressive strength of CO_3Ap-PLGA hybrid foam was as high as 0.4 MPa while CO_3Ap foam was only 0.01 MPa. Same concentration of PLGA solution without CO_3Ap particles was also reinforced on the CO_3Ap foam to check if the compressive strength would be different. No differences in compressive strength were found on CO_3Ap foam reinforced with PLGA only and PLGA mixed with CO_3Ap particles. Therefore, PLGA with low-crystalline CO_3Ap particles is a useful coating material to improve the compressive strength of CO_3Ap foam. The obtained compressive strength value was enough to keep the CO_3Ap-PLGA hybrid foam from crumbling when handled, thus useful for clinical application. Porosity of the CO_3Ap-PLGA hybrid foam on the otherhand was reduced to 85% compared to CO_3Ap foam, which has 95% porosity. Nevertheless, the pore structure and pore size was almost the same on both CO_3Ap-PLGA hybrid foam and CO_3Ap foam. These porous structures and pore interconnectivity are critical for cell migration, body fluid transport and invasion of new tissue ingrowth. In addition, these pores could provide more surface areas for osteoclast cells for resorption.

The low-crystallinity of CO_3Ap particles mixed with PLGA for the reinforcing solution also improved the bioactivity of the CO_3Ap foam. Based on XRD results, the crystallinity of CO_3Ap foam was higher than the CO_3Ap-PLGA hybrid foam. This was expected due to the preparation method. The CO_3Ap foam was prepared by hydrothermal treatment of αTCP foam at 150°C. On the otherhand, the CO_3Ap powder was synthesized at 37°C, which is the physiologic body temperature. Higher temperature treatment often results to higher crystallinity. Therefore, the CO_3Ap powder, which was synthesized at lower temperature, had the lower crystallinity.

Ideally, the original CO_3Ap foam should be prepared at low temperature to have low crystallinity. But the foam was prepared by using the so-called ceramics foam method to obtain a full three-dimensional interconnected porous structure. In the ceramics foam method, polyurethane foam is used as a template and then coated with ceramics slurry. High temperature is needed to burn out the template and to sinter the ceramics slurry since no compaction process is employed. CO_3Ap cannot be used as the starting material for the slurry because of its instability at high temperature. So αTCP was used as precursor because of its stability at high temperature and it's easy to be transformed to another phase by hydrothermal treatment. However, a minimum of 150°C for 24 hours was still needed to transform the αTCP foam to CO_3Ap foam. Fortunately, low-crystalline CO_3Ap powder was synthesized at physiologic body temperature and was mixed with PLGA as the reinforcing material. Although it was mixed with non-osteoconductive PLGA, the CO_3Ap powder was fully exposed to the outside environment such as the cells and body tissues than PLGA. Exposure of low-crystalline CO_3Ap may not only favor osteoblast but also resorption by osteoclast.

CONCLUSION

CO_3Ap-PLGA hybrid foam with improved mechanical strength and osteoconductivity was developed by reinforcing the CO_3Ap foam with a mixture solution of PLGA and low-crystalline CO_3Ap powder particles. It is expected that CO_3Ap-PLGA hybrid foam can be a useful material for bone substitution or scaffold for tissue engineering.

ACKNOWLEDGMENT

This study was supported in part by a Grant-in-aid for Scientific Research from the Ministry of Education, Culture, Sports, Science and Technology, Japan. The first author is grateful to Mitsubishi Corporation of Japan for the Scholarship grant.

REFERENCES

[1]B.S. Chang, C.K. Lee, K.S. Hong, H.J. Youn, H.S. Ryu, S.S. Chung, K.W. Park, Osteoconduction at porous hydroxyapatite with various pore configurations, *Biomaterials* **21**, 1291-1298 (2000)

[2]J.X. Lu, B. Flautre, K. Anselme, P. Hardouin, Role of interconnections in porous bioceramics on bone recolonization in vitro and in vivo, *J Mater Sci Mater Med* **10**, 111-120, (1999)

[3]R.Z. Legeros, Apatites in biological systems, *Prog Cryst Growth Charac* **4**, 1-45 (1981)

[4]R.Z. Legeros, Calcium phosphates in oral biology and medicine, *Monographs in oral sciences* **15**, (1991)

[5]Y. Doi, H. Iwanaga, T, Shibutani, Y. Moriwaki, Y. Iwayama, Osteoclastic responses to various calcium phosphates in cell cultures, *J Biomed Mater Res* **47**, 424-433 (1999)

[6]M. Hasegawa, Y. Doi, A. Uchida, Cell-mediated bioresorption of sintered carbonate apatite in rabbits, *J Bone Joint Surg (Br)* **85B**, 142-147 (2003)

[7]H. Wakae, A. Takeuchi, K. Udoh, S. Matsuya, M.L. Munar, R.Z. Legeros, A. Nakashima, K. Ishikawa, Fabrication of macroporous carbonate apatite foam by hydrothermal conversion of α-tricalcium phosphate in carbonate solutions, *J Biomed Mater Res A* **87**, 957-963 (2008)

[8]A. Takeuchi, M.L. Munar, H. Wakae, M. Maruta, S. Matsuya, K. Tsuru, K. Ishikawa, Effects of temperature on crystallinity of carbonate apatite foam prepared from α-tricalcium phosphate by hydrothermal treatment, *Biomed Mater Eng* **19**, 205-211 (2009)

UV-IRRADIATION MODIFIES CHEMISTRY OF ANATASE LAYER ASSOCIATED WITH IN VITRO APATITE NUCLEATION

Keita Uetsuki
R&D Center, Nakashima Medical Co. Ltd.
Haga, Kita-ku, Okayama 701-1221, Japan

Shinsuke Nakai, Yuki Shirosaki, Satoshi Hayakawa, Akiyoshi Osaka
Graduate School of Natural Science and Technology, Okayama University
Tsushima Naka, Kita-ku, Okayama 700-8530, Japan

ABSTRACT

Anatase layer was derived on pure titanium substrates by the H_2O_2 treatment and calcination, on which UV was irradiated in air (UVa) and in ultra pure water (UVw). The effects of the UV irradiation on in vitro apatite formation in Kokubo's simulated body fluid were interpreted in terms of the modified chemistry of the anatase layer. The UVa treatment reduced apatite formation while the contrary effects were observed for the UVw treatment. From O1s X-ray photoelectron spectroscopic (XPS) analysis, those results were correlated to the change in the relative amount of Ti-OH (basic TiOH) and Ti-O(H)-Ti (acidic Ti-OH). Combining the XPS analysis and the apatite growth characteristics, i.e., the number and size of the semi-spherical particles and their surface-coverage, derived a plausible model: proper assembly of those sites (Ti-OH and Ti-O(H)-Ti) that collect both calcium and phosphate ions together should only give rise to induction of apatite nucleation.

INTRODUCTION

The bio-inert characteristics of titanium and its alloys should be modified to attain better direct bonds to bone though they are commonly employed for orthopedic and dental implants because of their chemical, mechanical and biological properties. One of the modifications is to provide their surface with such an active titanium oxide layer that spontaneously deposits an apatite layer when they are in contact with plasma. This ability is sometimes denoted as "bioactivity," and the word is used even for the materials that deposit such apatite in vitro, like in the Kokubo's simulated body fluid (SBF; ISO 23317)[1]. Titanium oxide layers derived by chemical treatment with NaOH or H_2O_2[2] may be so modified by UV light irradiation as to be bioactive since Kasuga et al. [3] observed in vitro apatite formation on the UV-irradiated (rutile)$_{20}$-(anatase)$_{80}$ TiO_2 disc compacts after soaking them in 1.5 SBF for 5 to 30 d. They suggested that UV-irradiation induced the apatite nucleation. Shozui et al.[4] found that UV-irradiation provided bioactivity to the inert rutile surface layer derived on chemically pure titanium (cp-Ti) by calcination in air. Thus, UV-irradiation seems to be an effective strategy for improving bioactivity. Recently, we reported contrary effects of the UV irradiation on the bioactivity of specific anatase layer derived on cp-Ti substrates[5] due to CHT (Chemical and Heat Treatment) procedure[2]: the procedure was the combination of the chemical treatment with dilute H_2O_2 solution and subsequent heating in air. That is, the UV-irradiation in water (UVw) enhanced bioactivity of the CHT anatase layer but the UV-irradiation in air (UVa) reduced it.

In this study, we examined how the UV-irradiation under different environmental conditions chemically modified the CHT anatase layer. Several factors have been proposed to control the nucleation and growth of apatite such as negative charges and abundance in OH groups[6], crystal lattice matching between anatase or rutile or hydroxyapatite[7], hydrophilicity[8], and morphology[9]. As the CHT samples were used for a series of analyses in this study, most of those factors were ignorable that enabled to focus our attention to the surface chemistry, i.e., the effects of UVa and UVw treatments on the relative amount of certain Ti-OH or other OH-related species present on the anatase surface as well as their distribution.

EXPERIMENTAL

Mirror polished cp-Ti substrates (1 mm thick and 15 mm in diameter) were washed with acetone and ultra-pure water in an ultrasonic cleaner three times for 10 min at each step. Then, anatase-type TiO_2 layer was prepared due to the CHT procedure according to Wang et al.[2]: cp-Ti substrates were treated with 3% H_2O_2 solution at 80°C for 3h and then heated at 400°C for 1 h. The obtained samples were coded as CHT. Then, they were subsequently exposed to UV-light of a mercury lamp (HLR100T-2, SEN LIGHT Corp., Osaka, Japan; primary wavelength 365 nm, 170 mW/cm[2]) in air or in ultra-pure water for 1 h. The samples were coded as CHT_UVa and CHT_UVw, respectively.

SBF was prepared via the standard procedure and kept at 36.5°C with pH adjusted at 7.4[1,2,4,5]. After soaked in SBF up to 24 h, samples CHT, UVa and UVw were moderately rinsed with running ultra-pure water, dried at ambient temperature, and stored in a vacuum desiccator until they were subjected to the characterization.

The crystals deposited on the sample surfaces were identified using an X-ray diffractometer equipped with a thin-film attachment (TF-XRD, RINT2000, Rigaku, Tokyo, Japan; $CuK\alpha_1$, λ = 0.15406 nm, 40 kV, 200 mA). The incident beam was fixed at θ = 1°, and the detector was step-scanned around the 2θ axis from 20° to 40° at a rate of 0.01°/step with a count time of 1 s.

The surface morphology of the sample surface was examined by the field emission scanning electron microscopy (FE-SEM, S-4800, HITACHI High-Technology, Tokyo, Japan) after the samples were sputter-coated with a gold layer approximately 30 nm thick. From each FE-SEM image, the number and size of the deposited particles were obtained by image analysis software package (Image J, free software package from the National Institutes of Health) as a function of the period of soaking in SBF.

The samples were also subjected to O1s X-ray photoelectron spectroscopic analysis after soaked in acetone for 7 d (XPS; Fisons Instruments, S-prove ESCA SSX100S). The instrument was equipped with a monochromatic X-ray source (Al Kα), and the energy was presented in eV by convention. The same procedure was followed for the measurement as that of Hayakawa et al.[10], except that the binding energy drift was corrected by taking the measured O1s binding energy of the Ti2p core level to be 458.8 eV as the reference. The energy resolution of the XPS system was 0.16 eV. The spectra were simulated with a few component peaks for two OH groups and the lattice oxide ions according to Healy and Ducheyne[11].

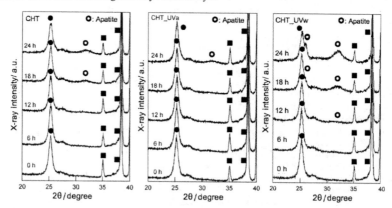

Fig. 1   TF-XRD profiles of CHT samples soaked in SBF. O: apatite; •: anatase; ■: Ti

RESULTS

Bioactivity or apatite forming ability

The TF-XRD patterns in Fig. 1 indicate that all samples CHT, CHT_UVa, and CHT_UVw gave almost equivalent diffraction profiles with the diffractions of anatase (JCPDS 21-1272) at

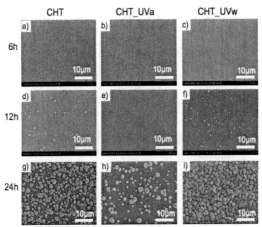

Fig. 2   SEM images of the surface microstructures of the CHT (a, d, g), CHT_UVa (b, e, h), and CHT_UVw (c, f, i) samples after soaked in SBF for up to 24 h.

25.3° and titanium (JCPDS 44-1294) at 35.1° and 38.4° in 2θ.   In addition, a new diffraction peak appeared at 25.9°, assignable to apatite[2-7], when the samples were soaked in SBF.   If the period until detectable TF-XRD intensity appeared was taken as the induction time for apatite nucleation τ_{ap}(xrd), τ_{ap} was around 12h for CHT_UVw < 18h for CHT < 24 h for CHT_UVa.   This was supported by their SEM images in Fig 2, demonstrating that many smaller particles were observed for CHT_UVw at 12h while a less number of particles was found for CHT, and none for CHT UVa.   From Fig. 2, the number density of the apatite particles at 12 h and 24 h increased in the order CHT UVa < CHT << CHT UVw.

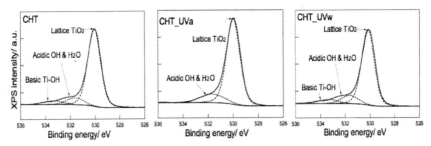

Fig. 3   _XPS spectra of the CHT samples with and without UV irradiation.   The deconvoluted component peaks were assigned after Healy & Ducheyne[12].

Surface chemical state
    Figure 3 shows the XPS O1s spectra, where the profiles were deconvoluted to a few component peaks[11].   The major peak at 530.1 eV was assigned to the oxide ions in the TiO_2 lattice.   One of the shoulder peaks at 531.6 eV was assigned to physically adsorbed water and oxygen atoms that were doubly coordinated to titanium (Ti-O(H)-Ti), and the other at 533.3 eV was attributed to chemically adsorbed H_2O and Ti-OH (Ti-OH) Basic OH).   The Ti-OH groups associated with the former and latter O1s peaks were denoted as acidic and basic Ti-OH, on the basis of the pKa values of 2.9 for Ti-O(H)-Ti and 12.7 for Ti-OH[12], respectively.   Note here that

the basic Ti-OH component disappeared due to the UVa treatment while both basic and acidic Ti-OH remained after the UVw treatment.

DISCUSSION

Process of apatite nucleation and growth

Crystallization kinetics involves two elementary reaction steps, nucleation and growth. τ_{ap}(xrd) is concerned with both heterogeneous nucleation on the active sites for apatite and growth of the nuclei since the XRD intensity is a measure of the amount of depositing apatite. A more precise way to derive τ_{ap}(xrd) is to extrapolate the X-ray intensity versus the soaking time profile to 0 intensity. This technique gave 6.5 h for CHT_UVw < 15 h for CHT < 19 h for CHT_UVa.

The number of apatite particles is directly correlated to the number of nuclei. Thus, Fig. 2 indicated that the UVa treatment suppressed the rate of nucleation while the UVw treatment accelerated the nucleation. That is, the surface layer of CHT_UVw had the largest number of active sites and CHT_UVa had the least. Careful observation of the apatite crystals in Fig. 2 indicates that each globular crystal was not a dense monolithic semisphere but an agglomerate of plate-shaped crystallites. This strongly suggests that even if each semispherical agglomerate was originated from a single nucleus, the hemisphere was the results of iterated secondary nucleation and growth of apatite on those platelet crystallites. In other words, once the nuclei were formed, they would grow in size on any sample in this study because each platelet had equivalent potential of repeating the secondary nucleation and growth under the same physiological environment, or SBF. Indeed, no significant difference was observed in the growth rate of the apatite particle size among the samples.

Chemical species responsible for heterogeneous nucleation

Despite that several factors[6-9] were proposed to control the in vitro apatite-forming ability of titania layer, certain changes in chemistry of the anatase surface should have affected the bioactivity or apatite formation. Hydrophilicity could not be a factor since all CHT samples with or without UV treatments were superhydrophilic and showed so low contact angle against with respect to water, <5°[2], that precise measurement was impossible. According to the O1s XPS analysis in Fig. 3, the most plausible factor is associated with the Ti-OH groups. The basic OH groups, generated by UV-irradiation[3,8,13] were frequently proposed to be an active site for apatite-nucleation and growth because the calcium ions in SBF were attracted to the negatively charged O-Ti[6,13] due to dissociation of the acidic Ti-OH. Although Kokubo et al.[14] recently proposed that positively charged Ti metal surface could also deposit apatite, majority described a similar scenario starting with attraction of the calcium ions by the negatively charged titania layer, admitting that the point of zero charge, p.z.c., of titania was 6.5, according to Boehm[12]. Those studies only emphasized the effects of total negative or positive charge on the material surface in SBF with pH 7.4. However, the valule pKa of basic OH (12.7) and acidic OH (2.9)[12] definitely allows coexistence of both negatively and positively charged species on the titania surface in SBF. Therefore, any of the simple models above in which negatively charged titania layer would attract positive ions is not applicable to interpret the apatite deposition.

From those discussions above, a working hypothesis is to be proposed that the active sites for inducing the primary apatite-nucleation in SBF should consist of positively and negatively charged chemical species associated with Ti-OH that are spatially arranged so as to collect the component species of apatite together in a well-organized manner. Fig. 4 depicts this hypothesis in a schematic way. That is, if the primary apatite-nucleation should depended on sufficient presence of such Ti-OH groups, the real factor to promote nucleation should be a way of adequate assembly of positively and negatively charged site. In other words, inadequate distribution of those active sites, however many there are, would not induce the primary apatite-nucleation on sample surface in SBF.

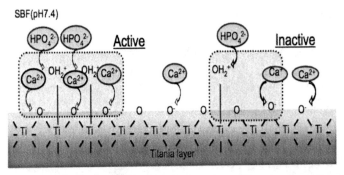

Fig. 4   A schematic diagram for a working hypothesis: active and inactive assemblies of Ti-OH groups for inducing apatite nucleation.   Nucleation needs both calcium and phosphate ions gathering around properly clustered Ti-OH groups.

CONCLUSION
    The effects of UV-irradiation on in vitro apatite formation on the CHT anatase layer were studied.   The number of active nucleation sites was increased by the UV irradiation on the CHT sample in water (CHT_UVw) while the contrary was found for the UV irradiation in air (CHT UVa), i.e., it was increased in the order CHT_UVa < CHT << CHT_UVw.   No practical change in the crystal growth rate was detected among those samples.   The XPS analysis indicated that the UVa treatment diminished basic OH with 12.7 in pKa while the UVw treatment caused little change in either basic OH or acidic OH with 2.7 in pKa.   From the results obtained, a hypothesis was proposed: the active sites for inducing primary apatite-nucleation should consist of spatially well-organized negatively and positively charged chemical species and the density and spatial arrangement of the chemical species should control the induction time and the rate for primary apatite nucleation by UV-irradiation.

ACKNOWLEDGMENT
    The authors thank Professor Tokuro NANBA at Okayama University for his assistance with the XPS measurements.   A part of this study was supported by the Kirameki Okayama Creation Fund and the Grant for Encouragement of Students, Graduated School of Natural Science and Technology, Okayama University.

REFERENCES
[1] T. Kokubo, T. Takadama. How useful is SBF in predicting *in vivo* bone bioactivity? *Biomaterials*, **27**, 2907-2915 (2006).
[2] X. X. Wang, S. Hayakawa, K. Tsuru, A. Osaka. A comparative study of *in vitro* apatite deposition on heat- H_2O_2-, and Na-OH treated titanium surfaces. *J. Biomed. Mat. Res.*, **54**, 172-178 (2001).
[3] T. Kasuga, H. Kondo, M. Nogami. Apatite formation on TiO_2 in simulated body fluid. *J. Cryst. Growth*, **235**, 235-240 (2002).
[4] T. Shozui, K. Tsuru, S. Hayakawa, A. Osaka. Enhancement of *in vitro* apatite-forming ability of thermally oxidized titanium surfaces by ultraviolet irradiation. *J. Ceram. Soc. Japan*, **116**, 530-535 (2007).
[5] K. Uetsuki, H. Kaneda, Y. Shirosaki, S. Hayakawa, A. Osaka. Effects of UV-irradiation on *in vitro* apatite-forming ability of TiO_2 layers. *Mat. Sci. Eng. B*, **173**, 213-215 (2010).
[6] P. Li, I. Kangasniemi, K. de Groot, T. Kokubo. Bonelike hydroxyapatite induction by a gel-derived titania on a titanium substrate. *J. Am. Ceram. Soc.*, **77**, 1307-1312 (1994).
[7] J. M. Wu, S. Hayakawa, K. Tsuru, A. Osaka. Low-temperature preparation of anatase and rutile

layers on titanium substrates and their ability to induce *in vitro* apatite deposition. *J. Am. Ceram. Soc.*, **87**, 1635-1642 (2004).

[8] Y. P. Guo, H. X. Tang, Y. Zhou, D.C. Jia, C. Q. Ning, Y. J. Guo. Effects of mesoporous structure and UV irradiation on *in vitro* bioactivity of titania coatings. *Appl. Surf. Sci.*, **16**, 4945-4952 (2010).

[9] J. Kunze, L. Müller, J. M. Macak, P. Greil, P. Schmuki, F. A. Müller. Time-dependent growth of biomimetic apatite on anodic TiO_2 nanotubes. *Electochemica. Acta*, **53**, 6995-7003 (2008).

[10] S. Hayakawa, A. Osaka, H. Nishioka, S. Matsumoto, Y. Miura. Structure of lead oxyfluorosilicate glasses: X-ray photoelectron and nuclear magnetic resonance spectroscopy and molecular dynamics simulation. *J. Non-Cryst. Solids.*, **272**, 103-118 (2000).

[11] K. E. Healy, P. Ducheyne. The mechanism of passive dissolution of titanium in a model physiological environment. *J. Biomed. Mat. Res.*, **26**, 319-338 (1992).

[12] H. P. Boehm. Acidic and basic properties of hydroxylated metal oxide surfaces. *Discuss. Faraday Soc.*, **52**, 264-275 (1971).

[13] Y. Han, D. Chen, J. Sun, Y. Zhang, K. Xu. UV-enhanced bioactivity and cell response of micro-arc oxidized titania coatings. *Acta Biomaterialia*, **4**, 1518-1529 (2008).

[14] T. Kokubo, D. K. Pattanayak, S. Yamaguchi, H. Takadama, T. Matsushita, T. Kawai, M. Takemoto, S. Fujibayashi, T. Nakamura. Positively charged bioactive Ti metal prepared by simple chemical and heat treatments, *J. Royal Soc. Interface*, **7**, 503-513 (2010).

PREPARATION OF MAGNESIUM CONTAINING BIOACTIVE TiO$_2$ CERAMIC LAYER ON TITANIUM BY HYDROTHERMAL TREATMENT

Xingling Shi, Masaharu Nakagawa, Lingli Xu and Alireza Valanezhad
Department of Biomaterials, Faculty of Dental Science, Kyushu University, JAPAN

ABSTRACT

Surface modification on titanium was carried out in order to improve its bioactivity. Pure titanium was hydrothermally treated in 0.1 mol/L MgCl$_2$ solutions with different pH at 200°C for 24hr. Surface morphology, roughness, wettability and chemical composition were characterized before and after treatment. Simulated body fluid (SBF) was employed to examine in vitro bioactivity. After hydrothermal treatment, nano-sized precipitations were observed and these surfaces showed superhydrophilicity. Ti-O-Mg chemical bonding was formed on titanium surface after hydrothermal treatment, however, under high pH, Mg(OH)$_2$ was precipitated from the solution. Hydrothermal treatment increased thickness of the oxide layer and the oxide layer was modified with magnesium (Mg). Apatite spheres were precipitated on hydrothermally treated samples after immersion in SBF for 2 days. However, the precipitation of apatite was depressed with increasing Mg concentration. The hydrothermal treatment was expected to be an effective method to fabricate titanium implant with good bioactivity.

INTRODUCTION

Titanium and its alloys are widely used for implants in dentistry and orthopedics. However, due to bioinertness, surface modifications are needed to improve their osteoconductivity.[1] The initial methods used for commercial products were simple physical treatments, such as sandblast and machining. Then, plasma spray hydroxylapatite (HAp) coating was widely researched. However, lack of long-term reliability due to low crystallinity limited its application.[2]

The excellent chemical inertness, corrosion resistance and good biocompatibility of titanium are thought to result from the oxide layer formed during processing. Furthermore, by observing apatite formation in vitro and in vivo TiO$_2$ film was realized also to play a vital role in the reactions between implant and bone.[3-5] Thus, TiO$_2$ films with various roughness, thickness and compositions were developed on Ti implants by different methods.[6-8]

Our previous studies showed that Ca can be immobilized into TiO$_2$ layer formed by hydrothermal treatment in Ca-containing solutions.[9,10] On the other hand, Mg plays essential roles in many physiological functions related to bone remodeling.[11-14] Therefore, in this study we investigated the surface properties of titanium modified by Mg through hydrothermal treatment.

EXPERIMENTAL

Hydrothermal treatments

Pure titanium disks (The Niraco Co., Tokyo, Japan) with diameter of 15 mm and thickness of 1 mm were used. All the samples were wet-abraded with SiC abrasive paper up to 1000# and then were ultrasonically washed in acetone and distilled water. Hydrothermal solutions were prepared by dissolving regent grade MgCl$_2$ (Wako Pure Chemical Industries, Ltd., Osaka, Japan) into distilled water with the concentration of 0.1 mol/L. In another experiment, 0.1 mol/L MgCl$_2$ solution with pH9.5 was prepared by adding diluted NaOH solution to study the effect of pH. 10 mL solution for each sample was filled into Teflon-lined vessel and samples were kept in an oven at 200°C for 24hr. After cooling down, samples were taken out, rinsed by distilled water and then washed ultrasonicaclly in distilled water for 10 min. Untreated samples and samples hydrothermally treated in distilled water (DW) at 200°C for 24hr were used as control. Untreated samples and samples hydrothermally treated in DW, MgCl$_2$ solution and high pH MgCl$_2$ solution are coded as UnTi, DWTi, MgTi and Na-MgTi, respectively.

Surface characterizations

Surface morphologies were observed with scanning electron microscope (SEM, JSM-5400LV, JEOL Ltd., Tokyo, Japan). Surface wettability was examined by contact angle meter (DM 500, Kyowa Interface Science Co., Ltd., Saitama, Japan). Surface chemical compositions were investigated by X-ray photoelectron spectroscopy (XPS, K-alpha, ThermoFisher Scientific, East Grinstead, UK). Crystal structure of the top layer was analyzed by X-ray diffraction (GIXRD, D8 Advance, Bruker AXS Inc., Germany).

Examination of apatite forming ability

After hydrothermal treatment, samples were immersed in 35 mL SBF at 36.5°C. The SBF was prepared by dissolving NaCl, NaHCO$_3$, KCl, K$_2$HPO$_3$·3H$_2$O, MgCl$_2$·6H$_2$O, CaCl$_2$, Na$_2$SO$_4$ into distilled water as proposed by Kokubo et al.[15] Samples were immersed in SBF to check the apatite formation.

RESULTS AND DISCUSSIONS

Surface morphology and wettability

No macroscopic difference was observed before and after hydrothermal treatment. However, SEM observation revealed that microstructure was different before and after the hydrothermal treatment. Scratches were observed on surface of abrasive paper polished sample. After the hydrothermal treatment nano-sized precipitates distribute all over the surface, whereas, topography of the scratches still remained. There was significant difference with respect to wettability before and after hydrothermal treatment. Titanium showed contact angle of 36° before the treatment. After the hydrothermal treatment, samples showed superhydrophilicity with contact angle approaching 0°.

Surface chemical composition

Surface chemical compositions investigation with XPS survey scan were showed in Fig. 1. Mg was combined into titanium surface, and higher pH benefited the immobilization. Sodium was not detected on Na-MgTi samples. According to binding energies, Mg on MgTi was thought to exist largely as magnesium titanate whereas Mg on Na-MgTi exist largely as magnesium hydroxide.

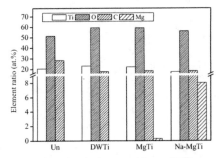

Figure 1. Surface chemical compositions of different samples

Depth profiles of oxide layers

Depth profiles of samples subjected to different treatments showed that the hydrothermal treatment increased the thickness of oxide layer obviously, as shown in Fig. 2. Mg immobilized into MgTi sample distributed uniformly through the oxide layer. However, for Na-MgTi sample, Mg concentrated on the top surface and its concentration decreased fast as the etching went on.

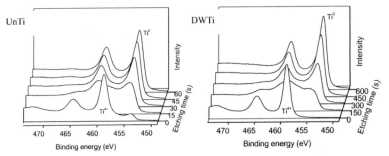

Figure 2. XPS Ti2p peaks of different samples after etching

In vitro apatite formation in SBF

When specimens were soaked in SBF at 36.5°C for 2d, deposits were observed only on hydrothermally treated samples as shown in Fig. 2. In other words, no deposit was observed on untreated samples. The deposits formed on hydrothermally treated Ti plate presented typical morphology of apatite precipitated from SBF. The composition was further confirmed by XRD. Samples treated in distilled water showed the best apatite inducing ability. Combination of Mg seemed to depress apatite formation. This may due to the fact that Mg inhibits nucleation of apatite and stabilizes amorphous calcium phosphate [16].

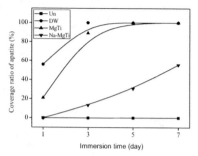

Figure 3. Coverage ratio of apatite on different samples depending on immersion time

CONCLUSION

Magnesium was successfully combined into TiO$_2$ formed by hydrothermal method in MgCl$_2$ solution. TiO$_2$ layer thickness was obviously increased by hydrothermal treatment and those samples showed good apatite inducing ability when immersed in SBF. However, Mg showed negative effect for the apatite inducing ability.

ACKNOWLEDGEMENTS

This study was supported in part by a grant-in-aid for scientific research from the Ministry of Education, Culture, Sports, Science, and Technology of Japan, as well as by the China Scholarship Council (CSC, No. 2009642007).

REFERENCES

[1]L. Le Guéhennec, A. Soueidan, P. Layrolle, and Y. Amouriq, Surface treatments of titanium dental implants for rapid osseointegration, Dent. Mater., 23, 844-54 (2007).

[2]K. Hayashi, T. Inadome, H. Tsumura, Y. Nakashima, and Y. Sugioka, Effect of surface roughness of hydroxyapatite-coated titanium on the bone-implant interface shear strength, Biomaterials, 15, 1187-91 (1994).

[3]L. L. Hench, Bioceramics - from Concept to Clinic, J. Am. Ceram. Soc., 74, 1487-510 (1991).

[4]T. Kokubo, Bioactive glass ceramics: properties and applications, Biomaterials, 12, 155-63 (1991).

[5]P. Li, C. Ohtsuki, T. Kokubo, K. Nakanishi, N. Soga, and K. de Groot, The role of hydrated silica, titania, and alumina in inducing apatite on implants, J. Biomed. Mater. Res., 28, 7-15 (1994).

[6]D. L. Cochran, P. V. Nummikoski, F. L. Higginbottom, J. S. Hermann, S. R. Makins, and D. Buser, Evaluation of an endosseous titanium implant with a sandblasted and acid-etched surface in the canine mandible: radiographic results, Clin. Oral Implants Res., 7, 240-52 (1996).

[7]X. L. Shi, L. L. Xu, and Q. L. Wang, Porous TiO(2) film prepared by micro-arc oxidation and its electrochemical behaviors in Hank's solution, Surf. Coat. Technol., 205, 1730-35 (2010).

[8]X. L. Shi, Q. L. Wang, F. S. Wang, and S. R. Ge, Effects of electrolytic concentration on properties of micro-arc film on Ti6Al4V alloy, Mining Science and Technology (China), 19, 220-24 (2009).

[9]M. Nakagawa and J. Yamazoe, Effect of CaCl(2) hydrothermal treatment on the bone bond strength and osteoconductivity of Ti-0.5Pt and Ti-6Al-4V-0.5Pt alloy implants, J. Mater. Sci. Mater. Med.,

(2009).

[10]M. Nakagawa, L. Zhang, K. Udoh, S. Matsuya, and K. Ishikawa, Effects of hydrothermal treatment with CaCl2 solution on surface property and cell response of titanium implants, *J. Mater. Sci.: Mater. Med.,* **16**, 985-91 (2005).

[11]L. A. Cohen L., Kitzes R., Bone magnesium, crystallinity index and state of body magnesium in subjects with senile osteoporosis, maturity-onset diabetes and women treated with contraceptive preparations, **2**, 70-75 (1983).

[12]Y. J. K. Liu C.C., Aloia J.F., Magnesium directly stimulates osteoblast proliferation, *J. Bone Miner. Res.,* **3**, 104-12 (1988).

[13]R. K. Rude, H. E. Gruber, L. Y. Wei, A. Frausto, and B. G. Mills, Magnesium deficiency: effect on bone and mineral metabolism in the mouse, *Calcif. Tissue Int.,* **72**, 32-41 (2003).

[14]H. Zreiqat, C. R. Howlett, A. Zannettino, P. Evans, G. Schulze-Tanzil, C. Knabe, and M. Shakibaei, Mechanisms of magnesium-stimulated adhesion of osteoblastic cells to commonly used orthopaedic implants, *J. Biomed. Mater. Res.,* **62**, 175-84 (2002).

[15]T. Kokubo and H. Takadama, How useful is SBF in predicting in vivo bone bioactivity?, *Biomaterials,* **27**, 2907-15 (2006).

[16]N. Sahai, Modeling apatite nucleation in the human body and in the geochemical environment, *Am. J. Sci.,* **305**, 661-72 (2005).

MILLIMETER-SIZED GRANULES OF BRUSHITE AND OCTACALCIUM PHOSPHATE FROM MARBLE GRANULES

A. Cuneyt Tas
Department of Biomedical Engineering, Yeditepe University
Istanbul 34755, Turkey

ABSTRACT
Brushite (DCPD, dicalcium phosphate dihydrate, $CaHPO_4 \cdot 2H_2O$) and octacalcium phosphate (OCP, $Ca_8(HPO_4)_2(PO_4)_4 \cdot 5H_2O$) granules with diameters in the millimeter range were prepared by using marble (calcium carbonate) granules as the starting material. The method of this study simply comprised of soaking the marble granules in aqueous solutions containing phosphate and/or calcium ions at temperatures between 20° and 37°C. This process did not cause any size change between the initial marble and final brushite or octacalcium phosphate granules. Such DCPD and OCP granules of higher in vitro solubility than hydroxyapatite could be useful in maxillofacial, dental and orthopedic void/bone defect filling and grafting applications. Samples were characterized by X-ray diffraction, inductively-coupled plasma atomic emission spectroscopy, and scanning electron microscopy.

INTRODUCTION
It is easy to synthesize powders of brushite (DCPD, dicalcium phosphate dihydrate, $CaHPO_4 \cdot 2H_2O$) [1-5] and octacalcium phosphate (OCP, $Ca_8(HPO_4)_2(PO_4)_4 \cdot 5H_2O$) [6-10]. The log K_{SP} (solubility product) values for DCPD, OCP and hydroxyapatite (HA, $Ca_{10}(PO_4)_6(OH)_2$) are known to be -6.6, -96.6 and -117.1, respectively [11]. Therefore, one may expect solely on the basis of the above solubility values that both DCPD and OCP shall exhibit in vivo percentages of biodegradation and bone replacement higher than that of hydroxyapatite.
DCPD is an acidic calcium phosphate and when implanted in the form of a cement paste containing some β-TCP (β-$Ca_3(PO_4)_2$) granules, the material was initially surrounded by macrophages and fibrous tissue due to the initial pH drop, but still showed a higher biodegradation rate than that of a control hydroxyapatite cement over the long run [12, 13]. Unfortunately, in vitro and in vivo data on pure DCPD powders are lacking to the best of our knowledge. On the other hand, to the advantage of OCP, when it was implanted as a dry-sieved powder (sieving apparently causing a certain degree of aggregation of the particulates) such fibrous tissue encapsulation was not observed, and OCP biodegraded well [14].
The difficulty arises if one wants to produce granules or bodies of DCPD or OCP, instead of powders, which should have at least the strength to withstand the handling by the surgeon during implantation. DCPD or OCP cannot be sintered, in stark contrast to HA or β-TCP bioceramics, mainly because at temperatures above 60°C they start to transform into $CaHPO_4$ (DCPA, monetite, dicalcium phosphate anhydrous) and HA, respectively. In the literature, there were only scarce attempts to prepare granules of monetite [15-17]. The starting materials in these studies were either brushite cement paste containing β-TCP [15, 16] or pure β-TCP [17]. The so-called brushite cement pastes are prepared by reacting orthophosphoric or sulphuric acid (cement setting solution) with β-TCP (69.74 wt%), $Ca(H_2PO_4)_2 \cdot H_2O$ (30 wt%), and $Na_2P_2O_7$ (<0.26 wt%), therefore, such pastes contained a significant of amount unreacted β-TCP in their cores [16, 18].
The down-to-earth approach taken in this study was to use granules of natural marble (calcite, $CaCO_3$) as the starting material and transforming them firstly into DCPD and secondly to OCP by soaking in specially prepared aqueous solutions in glass media bottles (without stirring) at temperatures between 20° (room temperature, RT) and 37°C. The sizes of the obtained DCPD and OCP granules closely imitated the initial sizes of the marble granules used.

EXPERIMENTAL PROCEDURE

Marble granules were purchased from Merck KGaA, Darmstadt, Germany, (Cat. No: 105986). The granules were 0.9 to 2 mm in size. Marble granules were used as-received, without any further purification or chemical treatment, only after dry sieving manually by using a sieve with 0.9 mm openings to remove very small amounts of marble dust that could be present.

For the preparation of DCPD granules, ten grams of $NH_4H_2PO_4$ (Cat. No: 101126, Merck) was dissolved in 50 mL of doubly distilled water. The pH of the obtained solution was 4.1 ±0.1 at RT. The solutions were prepared in 100 mL-capacity glass media bottles. In some experiments, that 10 g of $NH_4H_2PO_4$ was replaced with 13.557 g of $NaH_2PO_4 \cdot 2H_2O$ (Cat. No: 106342, Merck) or 11.826 g of KH_2PO_4 (Cat. No: 104877, Merck). Two grams of marble granules were then placed into the phosphate solution. The granules were kept in sealed glass bottles at RT for about 20 h without stirring. Granules were separated from the solution by using a small sieve with 0.9 mm openings, followed by washing with 1.5 L of distilled water and overnight drying at 37°C.

An alternative phosphate solution was also prepared, for transforming marble granules into DCPD, by slowly adding 120 mL of concentrated (85%) H_3PO_4 (Cat. No: 100573, Merck) to 730 mL of doubly distilled water, followed by the drop-wise addition of 137 mL of concentrated (28-30%) NH_4OH (Cat. No: 105423, Merck); the resultant solution again had a pH value equal to 4.1 ±0.1 at RT. One mL of this solution had 1.889×10^{-3} mole P. In using this alternative solution to form the DCPD granules, 3 g of marble granules were placed into 69 mL of the above solution and were kept unstirred for 20 h at RT.

In synthesizing the OCP granules, DCPD granules as produced above were used as the starting material. The OCP synthesis solution was prepared by adding the following chemicals at RT, one by one, to 1420 mL of double distilled water under vigorous stirring; 12.448 g of NaCl (Cat. No: 106404, Merck), 0.559 g KCl (Cat. No: 104933, Merck), 0.426 g Na_2HPO_4 (Cat. No: 106586, Merck) and 0.735 g $CaCl_2 \cdot 2H_2O$ (Cat. No: 102382, Merck). 8.5 g of Tris (Cat. No: 108382, $(HOCH_2)_3CNH_2$, Merck) was then added to this slightly turbid solution, followed by titrating the solution to pH 7.4±0.2 by adding 60 to 67 mL of 1 mol/L HCl solution (Cat. No: 109057, Merck). The details of preparing the above solution are given in Table 1.

Table I   Preparation of the solution used in converting DCPD granules into OCP granules

Reagent	Order	Amount (g)
H_2O	1	1420
NaCl	2	12.448
KCl	3	0.559
Na_2HPO_4	4	0.426
$CaCl_2 \cdot 2H_2O$	5	0.735
Tris	6	8.500
1 M HCl	7	60 to 67 mL (*to adjust the pH at 7.4 at RT*)

To produce OCP granules, 500 mL of this transparent solution (with a Ca/P molar ratio of 1.667) was placed into a 500 mL-capacity glass media bottle and 2.2 g of DCPD granules were added into the bottle. The bottle was kept undisturbed at 37°C in a microprocessor-controlled oven for one week, however the solution in the bottle was fully replenished at the end of the first 3 days. This solution (of Table 1) was inspired from the one described by Wen et al. [19]. Granules were finally separated from the solution by using a small sieve with 0.9 mm openings, followed by washing with 1.5 L of distilled water and overnight drying at 37°C.

All granules were characterized by using an X-ray diffractometer (XRD; Advance D8, Bruker AG, Karlsruhe, Germany) after either gently mounting the granules in the sample holder on modeling clay (Play-Doh®) or after grinding them into a powder by using an agate mortar and pestle. XRD was operated with a Cu tube at 40 kV and 40 mA equipped with a monochromator. Samples were scanned with a step size of 0.02° and a preset time of 5 s.

Scanning electron microscopy (SEM; EVO 40, Zeiss, Dresden, Germany) was used to evaluate the morphology of the granules, after sputter-coating the granules with a 25 nm-thick gold layer to impart electrical conductivity to the specimen surfaces.

Chemical analyses of granules were performed by using inductively-coupled plasma atomic emission spectroscopy (ICP-AES; Model 61E, Thermo Electron, Madison, WI). For the ICP-AES analyses, roughly 100 mg portions of granules were dissolved in 7.5 mL of concentrated (69%) HNO_3 (Cat. No: 101799, Merck) solution.

RESULTS AND DISCUSSION

The chemical analyses of the initial marble granules were performed by using ICP-AES and the granules were found to consist of 55.5 wt% CaO, 2100 ppm MgO, 960 ppm SiO_2, 430 ppm Al_2O_3, and <200 ppm Fe_2O_3. The values above were the average of three measurements. No other impurities were detected. X-ray diffraction (XRD) analyses of the marble granules showed that they were consisting of single-phase calcite of high crystallinity, conforming very well with the standard powder diffraction file (PDF) of 5-0586 of International Centre for Diffraction Data (ICDD). The (104) reflection of the pristine marble granules, observed at 29.4° 2θ, was having an X-ray diffraction intensity higher than 15,000 counts-per-second. Figure 1 depicted the SEM morphology of the initial, white marble granules, with the inset showing a digital camera macrograph of the same; incorporating a millimetric scale to facilitate the direct evaluation of granule sizes.

Figure 1 SEM micrograph of the starting marble granules (*inset: optical macrograph of the same*)

Upon soaking the marble granules in the pure ammonium dihydrogen phosphate solution for 20 h at RT, the granules were covered with characteristic DCPD crystals (Figures 2a and 2b). The SEM morphology did not exhibit any noticeable change when the marble granules were similarly soaked in the alternative $NaH_2PO_4 \cdot 2H_2O$, KH_2PO_4 or H_3PO_4-NH_4OH solutions (as described in Chapter 2) and all four liquid media were able to transform the marble granules into DCPD. The solution pH increased to the range of 5.1 to 5.3 at the end of 20 h of ageing. The XRD pattern of the DCPD granules was given in Figure 3, all the peaks belonging to the DCPD phase (peak positions matching perfectly well with those given in ICDD PDF 09-0077). DCPD granules obtained by using $NaH_2PO_4 \cdot 2H_2O$, KH_2PO_4 and H_3PO_4-NH_4OH solutions also had the same XRD data (not shown). The XRD data was collected from the DCPD granules without grinding them into a powder (after mounting the granules directly in the sample holder). This was the reason why no $CaCO_3$ peak was observed in this figure. The high intensity (020) reflection located at 11.68° 2θ was characteristic of DCPD. One distinctive feature of these DCPD granules was that they were shining especially under the direct sun light due to the brushite crystals shown in Figures 2a and 2b. It was actually impossible to obtain any calcium phosphate phase other than DCPD when immersing, at RT, such a reactive calcium carbonate/marble material in a pure phosphate solution having a pH value of 4 [2, 8].

0.4 mm

Figure 2a   SEM micrograph of DCPD granules (*low magnification*)

Figure 2b   SEM micrograph of DCPD granules (*high magnification*)

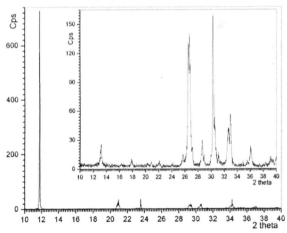

Figure 3   XRD chart of DCPD granules obtained by using the $NH_4H_2PO_4$ solution at RT, no stirring   (*inset: XRD of DCPA powders obtained after stirring the marble granules in the $NH_4H_2PO_4$ solution for 20 h at 65°C*).

ICP-AES analyses of the DCPD granules showed that the Na and K amounts were less than 100 ppm, even when we used $NaH_2PO_4 \cdot 2H_2O$ or KH_2PO_4 solutions. If one added, for instance, precipitated chalk (i.e., calcite) powder into the same solution and stirred it vigorously at RT for 20 h, one would have obtained brushite powder. Similarly, when we added the same amount of marble granules into the same volume of the ammonium phosphate solution in a sealed glass bottle and stirred the solution at 65°C (at 400 rpm) constantly for 20 h, we obtained only a powder (no granules left) consisting of single-phase monetite ($CaHPO_4$), as shown in the XRD data of inset of Figure 3, where peak positions and intensities matched with those of ICDD PDF 09-0080. Increased temperature and vigorous stirring both forced (i) the DCPD to transform into DCPA and (ii) the marble granules to crumble into a powder.

It would be very difficult, if not impossible, to obtain OCP (octacalcium phosphate) directly from marble in an unstirred hydrothermal ageing solution of pH 7.4. At that pH, apatitic CaP is the phase to form. This was why we preferred DCPD granules as the starting material in producing the OCP granules. On the other hand, it would have been very easy to convert the DCPD granules into apatitic CaP granules by soaking them in a Ca^{2+}-containing solution with pH on the basic side (e.g., 8.5 to 10), but our intention here was not to produce low solubility HA. The solution developed [19] for the conversion of DCPD to OCP had a Ca/P molar ratio of 1.667 and the mildly acidic DCPD granules (with Ca/P ratios roughly equal to 1, depending on the amount of $CaCO_3$ remaining still unreacted in the cores of those granules) would help to slowly lower both the nominal Ca/P ratio in the ageing bottles to around 1.33 (i.e., to that of OCP) and the solution pH to around 6.5 to 6.8 (stability range of OCP) [9]. The DCPD-to-OCP transformation was a surface reaction.

Figures 4a and 4b depicted both the XRD data and the SEM micrographs of the OCP granules obtained by using the DCPD granules prepared in $NH_4H_2PO_4$ solutions. OCP granules produced by using the DCPD granules obtained in the other solutions of this study also showed identical XRD and SEM data. The peak positions and intensities of the XRD pattern of the OCP (without grinding the granules into a powder, Fig. 4a) granules were in good match with that of ICDD PDF 26-1056. The intermingling nanocrystals seen in Figure 4b depicted the characteristic habit of OCP crystals [10].

The OCP granule synthesis process described here is a benign one since it starts with a solution of neutral pH and proceeds at the human body temperature. The solution pH values were measured to be in the range of 6.5 to 6.8 at the end of ageing periods.

The overall conversion of brushite to OCP can be visualized by the below reactions. Reactions help to explain the effect of Ca^{2+} and HPO_4^{2-} ions present in the ageing solutions, as well as the slight pH decreases observed.

$$6CaHPO_4 \cdot 2H_2O(s) + 2Ca^{2+} \rightarrow Ca_8(HPO_4)_2(PO_4)_4 \cdot 5H_2O(s) + 7H_2O + 4H^+ \qquad (1)$$

$$5CaHPO_4 \cdot 2H_2O(s) + 3Ca^{2+} + HPO_4^{2-} \rightarrow Ca_8(HPO_4)_2(PO_4)_4 \cdot 5H_2O(s) + 5H_2O + 4H^+ \qquad (2)$$

Mg^{2+} ions are known [20] to significantly inhibit the nucleation and growth of cryptocrystalline apatitic calcium phosphate in aqueous media, and for this reason the presence of magnesium in the original marble granules were deemed to be simply advantageous for the synthesis of OCP. We did not assess the influence of other elements, such as Si, Al and Fe, present in smaller amounts in the initial marble granules.

Upon grinding (with an agate mortar/pestle) both the so-called DCPD and OCP granules of this study into a fine powder and then collecting their XRD data, the small amounts of $CaCO_3$ remaining in the cores of the granules became visible, as shown in Figure 5a. This was quite an expected observation, if there were no calcite remaining in the cores, the granules would have turned into powder by the end of the hydrothermal ageing process.

Figure 4a   XRD and SEM data of OCP granules

Figure 4b   High magnification SEM photomicrograph of OCP granules showing surface nanotexture

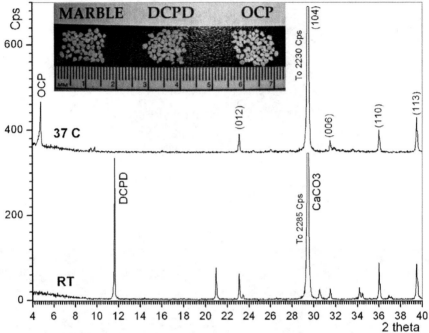

Figure 5a   XRD charts of DCPD (*bottom*) and OCP (*top*) granules (*after grinding into a fine powder*); reflections of calcite are indicated by their crystallographic planes (104) and observed intensities   *(inset shows the size similarity between the granules)*

The strongest (104) reflections of calcite located at 29.4° 2θ in both DCPD and OCP samples (Fig. 5a) were still much lower than the 15,000 Cps intensity of the initial marble granules. The percentage decrease in the XRD intensity of the (104) reflection of calcite (from the initial marble granules (15,000 Cps) to the final DCPD or OCP granules (2300 Cps)) corresponded to about 85%. Therefore, the process reported here was starting at the surfaces of the marble granules, proceeding inwards and then stopping at one point where the formed DCPD or OCP layers were partially obstructing the reach of the solution to the remnant calcitic marble cores at RT and 37°C.

What would happen if only the temperature and ageing time was increased, without stirring, in preparing the DCPD granules? The bottom trace of Figure 5b showed the XRD pattern of granules heated at 50°C for 96 h without stirring, whereas the top trace showed the granules heated at 70°C for 96 h again without stirring. In both cases, the end products were intact granules, not powders; in contrast to the stirred sample of Figure 3. It must be noted that while the intensity of the (104) reflection of the calcite phase was decreasing from 2285 Cps (RT, 20 h) to 1942 Cps (50°C, 96 h), upon increasing the solution ageing temperature to 70°C, DCPD phase completely decomposed to DCPA with a further decrease in the intensity of the calcite peak (Fig. 5b, top trace).

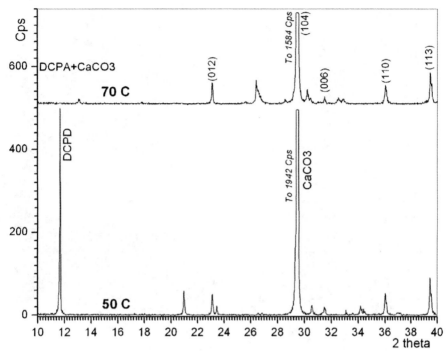

Figure 5b XRD charts of marble granules heated at 50°C for 96 h (*bottom*) and 70°C for 96 h (*top*) in the NH₄H₂PO₄ solution without stirring (*after grinding the granules into a fine powder*)

DCPD is a low-temperature phase. The reason behind the decrease in the XRD intensity of the remnant calcite in the cores of the granules was most probably the reduced surface tension of the aqueous ageing solution (in going from RT to 50°C and 70°C), and thus the solution was able to penetrate more inwards through the DCPD layer and reaching a bit more of the non-reacted, shielded marble at the granule cores. Increasing the ageing time also helped this transformation.

We further pursued this high-temperature transformation of marble granules by adding 2.35 g of marble granules to 140 mL of the ammonium phosphate solution (prepared by mixing concentrated solutions of H_3PO_4 and NH_4OH, see Chapter 2) in a non-stirred pressure vessel (Model 4760, Parr Instrument Company, Moline, IL) with a Teflon® cup inside, as shown in Fig. 6. The pressure vessel was heated in a microprocessor-controlled oven. By this way, we were able to reach 200°C (and stay there for 24 h) without any risk of evaporating the solution. This pressure vessel was not equipped with a pressure gauge.

Figure 6   Pressure vessel assembly used in testing the feasibility of forming granules between 90° and 200°C

The XRD traces of marble granule samples heated at temperatures between 90°C (96 h) and 200°C (24 h) are given in Figure 7. None of these samples disintegrated into a powder, they were all recovered as strong granules. Starting from somewhere between 90° and 110°C, a new phase was appearing: β-TCP. Heating runs at 150° and 200°C, on the other hand, also resulted in another phase: HA. Therefore, as shown in Figure 7, the samples heated in the pressure vessel at 110°, 150° and 200°C crystallized β-TCP besides DCPA, and the samples heated at 150° and 200°C were consisting of four phases: DCPA, β-TCP, HA, and calcite. The appearance of HA and β-TCP phases should be at the expense of calcite and DCPA. Crystallization of β-TCP at a temperature as low as 110°C was noteworthy [21]. The decrease observed in the amount of calcite (Fig. 7), with an increase in the temperature of the pressure vessel from 150° to 200°C, implied that a further increase in temperature up above 200°C could have totally consumed the calcite phase. Nevertheless, to prepare granules comprising the phases of HA and β-TCP was not one of our initial purposes, and for this reason we opt not to deliberate much on these multi-phase high temperature-high pressure samples.

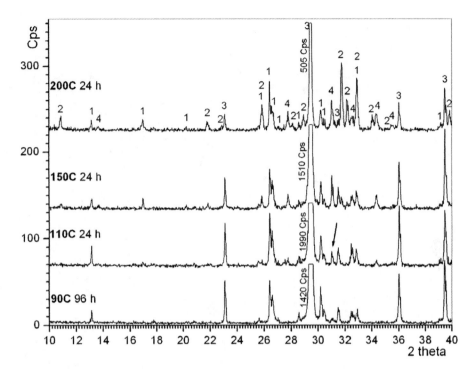

Figure 7   XRD traces of marble granules heated in a pressure vessel between 90° and 200°C, in a solution prepared by adding NH$_4$OH into H$_3$PO$_4$.
Phases = 1: DCPA, 2: HA, 3: calcite, 4: β-TCP.

The DCPD and OCP granules of this study may be expected to show a high *in vivo* degradability since their cores were not made out of β-TCP (log K_{SP}= -81.7 [11]) as was the case with the previous granule synthesis studies [16-18], but were out of calcite, CaCO$_3$ (log K_{SP}= -8.55 [11]). Calcite is more soluble than β-TCP by about an order of magnitude, and this fact was also proved by the *in vitro* cell culture study of Monchau *et al.* [22], which directly compared HA, β-TCP and CaCO$_3$ with one another by using both human and rat osteoclasts.

The preparation of biphasic DCPD-OCP granules (in any desired weight percentages) would be as easy as weighing and mixing the DCPD (log K_{SP}=-6.6) and OCP (log K_{SP}=-72.5) granules with one another. Such biphasic granule mixtures may thus provide a tool of fine-tuning the desired *in vivo* solubility/resorbability level, between the solubilities of the end members DCPD and OCP, of these biomaterials.

It is obvious that the weakest part of this *proof-of-the-concept* study was the absence of compressive strength measurements on the marble, DCPD and OCP granules (although it was still impossible to crush the DCPD and OCP granules into a powder by using the fingertips), as well as the *in vitro* and *in vivo* characterization, however, our follow-up study is already focusing right on those aspects.

CONCLUSIONS
Marble (calcite, $CaCO_3$) granules over the size range of 0.9 to 2 mm were used as the starting material to prepare brushite (DCPD, $CaHPO_4 \cdot 2H_2O$) granules, without changing the starting granule sizes. DCPD granule synthesis consisted of soaking the marble granules at room temperature in an acidic (of pH 4) phosphate solution, without stirring, for less than a day. DCPD granules were found to contain small amounts of residual marble in their cores. DCPD granules were then used as the starting material for the preparation of octacalcium phosphate (OCP, $Ca_8(HPO_4)_2(PO_4)_4 \cdot 5H_2O$) granules. A non-stirred Tris-buffered solution (of pH 7.4) containing Na^+, K^+, Ca^{2+}, Cl^- and HPO_4^{2-} ions was used in synthesizing OCP granules at 37°C in one week. DCPD and OCP granules were found to contain a small amount of residual marble ($CaCO_3$, calcite) in their cores.

Notes
Certain commercial instruments or materials are identified in this paper to foster understanding. Such identification does not imply recommendation or endorsement by the author, nor does it imply that the instruments or materials identified are necessarily the best available for the purpose.

REFERENCES
[1] H. Monma and T. Kamiya, "Preparation of Hydroxyapatite by the Hydrolysis of Brushite," *J. Mater. Sci.*, **22** 4247-4250 (1987).
[2] R.I. Martin and P.W. Brown, "Phase Equilibria Among Acid Calcium Phosphates," *J. Am. Ceram. Soc.*, **80** 1263-6 (1997).
[3] G.R. Sivakumar, E.K. Girija, S.N. Kalkura, and C. Subramanian, "Crystallization and Characterization of Calcium Phosphates: Brushite and Monetite," *Cryst. Res. Tech.*, **33** 197-205 (1998).
[4] R. Tang, C.A. Orme, and G.H. Nancollas, "A New Understanding of Demineralization: The Dynamics of Brushite Dissolution," *J. Phys. Chem. B.*, **107** 10653-7 (2003).
[5] S. Mandel and A.C. Tas, "Brushite to Octacalcium Phosphate Transformation in DMEM Solutions at 36.5°C," *Mater. Sci. Eng. C*, **30** 245-254 (2010).
[6] W.E. Brown, J.P. Smith, J.R. Lehr, and A.W. Frazier, "Crystallographic and Chemical Relations between Octacalcium Phosphate and Hydroxyapatite," *Nature*, **196** 1048-50 (1962).
[7] R.Z. LeGeros, "Preparation of Octacalcium Phosphate (OCP): A Direct Fast Method," *Calcified Tissue Int.*, **37** 194-7 (1985).
[8] G.H. Nancollas, M. Lore, L. Perez, C. Richardson, and S.J. Zawacki, "Mineral Phases of Calcium Phosphate," *Anat. Rec.*, **224** 234-41 (1989).
[9] M. Iijima, "Formation of Octacalcium Phosphate in vitro," in *Octacalcium Phosphate*. L.C. Chow and E. D. Eanes (eds.), Monogr. Oral Sci., Vol. 18, pp. 17-49, Karger, Basel, 2001.
[10] O. Suzuki, "Octacalcium Phosphate: Osteoconductivity and Crystal Chemistry," *Acta Biomater.*, **6** 3379-87 (2010).
[11] F.C.M. Driessens and R.M.H. Verbeeck, *Biominerals*, pp. 37-59, CRC Press, Boca Raton, FL, 1990.
[12] J.M. Kuemmerle, A. Oberle, C. Oechslin, M. Bohner, C. Frei, I. Boecken, and B. von Rechenberg, "Assessment of the Suitability of A New Brushite Calcium Phosphate Cement for Cranioplasty – An Experimental Study in Sheep," *J. Cranio Maxill. Surg.*, **33** 37-44 (2005).
[13] F. Theiss, D.Apelt, B. Brand, A. Kutter, K. Zlinszky, M. Bohner, S. Matter, C. Frei, J.A. auer, and B. von Rechenberg, "Biocompatibility and Resorption of a Brushite Calcium Phosphate Cement," *Biomaterials*, **26** 4383-94 (2005).

[14] O. Suzuki, S. Kamakura, T. Katagiri, M. Nakamura, B. Zhao, Y. Honda, and R. Kamijo, "Bone Formation Enhanced by Implanted Octacalcium Phosphate involving Conversion into Ca-deficient Hydroxyapatite," *Biomaterials*, **27** 2671-81 (2006).

[15] F.T. Marino, J. Torres, I. Tresguerres, L.B. Jerez, and E.L. Cabarcos, "Vertical Bone Augmentation with Granulated Brushite Cement Set in Glycolic Acid," *J. Biomed. Mater. Res.*, **81A** 93-102 (2007).

[16] F. Tamimi, J. Torres, C. Kathan, R. Baca, C. Clemente, L. Blanco, and E.L. Cabarcos, "Bone Regeneration in Rabbit Calvaria with Novel Monetite Granules," *J. Biomed. Mater. Res.*, **87A** 980-5 (2008).

[17] L.G. Galea, M. Bohner, J. Lemaitre, T. Kohler, and R. Mueller, "Bone Substitute: Transforming β-Tricalcium Phosphate Porous Scaffolds into Monetite," *Biomaterials*, **29** 3400-7 (2008).

[18] B. Flautre, C. Maynou, J. Lemaitre, P. van Landuyt, andP. Hardouin, "Bone-Colonization of β-TCP Granules Incorporated in Brushite Cements," *J. Biomed. Mater. Res.*, **63B** 413-7 (2002).

[19] H.B. Wen, J.G.C. Wolke, J.R. de Wijn, Q. Liu, F.Z. Cui, and K. de Groot, "Fast Precipitation of Calcium Phosphate Layers on Titanium Induced by Simple Chemical Treatments," *Biomaterials*, **18** 1471-8 (1997).

[20] F. Barrere, C.A. van Blitterswijk, K. de Groot, and P. Layrolle, "Nucleation of Biomimetic Ca-P Coatings on Ti6Al4V from a SBFx5 Solution: Influence of Magnesium," *Biomaterials*, **23** 2211-20 (2002).

[21] S. Jalota, A. C. Tas, and S. B. Bhaduri, "Microwave-assisted Synthesis of Calcium Phosphate Nanowhiskers," *J. Mater. Res.*, **19** 1876-1881 (2004).

[22] F. Monchau, A. Lefevre, M. Descamps, A. Belquin-Myrdycz, P. Laffargue, and H.F. Hildebrand, "In Vitro Studies of Human and Rat Osteoclast Activity on Hydroxyapatite, β-Tricalcium Phosphate, Calcium Carbonate," *Biomol. Eng.*, **19** 143–52 (2002).

MICROSTRUCTURES AND PHYSICAL PROPERTIES OF BIOMORPHIC SISIC CERAMICS
MANUFACTURED VIA LSI-TECHNIQUE

Steffen Weber, Raouf Jemmali, Dietmar Koch and Heinz Voggenreiter
Department of Ceramic Composite and Structures, Institute of Structures and Design
German Aerospace Center, Stuttgart, Germany

ABSTRACT
    Biomorphic SiSiC ceramic composites were produced by liquid silicon infiltration (LSI)
process. These materials are based on activated coal (ACBC) and wood fibers (MDF). Microstructural
characteristics in green-, carbon- and SiSiC-body modification were investigated by analyzing curved
structures (type charcoal) and planar structures (type MDF). Therefore polished and unpolished
specimens have been examined with scanning electron microscopy (SEM), microfocus X-ray
computed tomography (CT) and respective evaluation software. Focused on the objective of near net
shape reproducibility for the ceramic composites, different shrinkage and silicon uptake behavior were
identified. Furthermore, significant capillarities, densities and porosities could be determined and
valued regarding the objective and highest possible SiC content. The appearing different SiSiC
microstructures influence the physical properties of SiSiC ceramics such as flexural strength, Young's
modulus and hardness, which were detected with standardized testing methods.

INTRODUCTION
    New innovated materials based on biogenic raw materials are becoming increasingly important
in applied CMC technology. Biomorphic SiSiC ceramics, produced by liquid silicon infiltration
process (LSI)[1,2], offer an alternative to commercial sintered silicon carbide ceramics. Consequently,
the use of expensive ceramic powders or granulates and also the necessary high energy level by
sintering can be avoided. Due to the high stiffness as well as hardness, good corrosion and temperature
resistance biomorphic SiSiC ceramics provides an opportunity in the application sectors energy
technology, commercial mechanical engineering and ballistic protection.[3-8] Thereby, manufacture of
complex shaped structures with application-specific microstructural properties become the focus of
development. In addition, the possibility of near net-shape production of complex shaped SiSiC
structures marks a further key factor for economic use in various applications.[9,10] In this context,
shrinkage is a problem of wood based biomorphic materials during liquid silicon infiltration process
steps. However, activated coal based compounds offer strongly reduced shrinkage and therefore net
shape manufacturing of complex shaped ceramic structures. Furthermore, variable shaping options
arising from the use of warm pressing process step in the early beginning of LSI-process.
    The aim of this paper was the analysis of basic physical properties and generating more
information about the configurations and their characteristics after significant processing steps in LSI-
route of biomorphic SiSiC ceramics based on activated coal and wood fibers. These investigations
were focused on the characterization of pore channel systems and microstructures. In addition, first
information about processing curved structures by powdery compounds via warm pressing and
subsequent silicon infiltration could be provided.

EXPERIMENTAL PROCEDURE

Material composition and manufacture
    The produced and investigated biomorphic SiSiC ceramics are based on different raw material
composition. A distinction is made in wood ceramic type MDF, results from commercially available
medium density fiberboard and ceramics based on activated carbon type ACBC. In the case of ACBC
ceramics a phenolic resin is used as a binder and additional carbon source. Two activated carbon

variants with different surface characteristics were used. As a further extension, two material grades are reinforced with carbon fibers. The compounds based on carbon were produced by mixing and simultaneously homogenizing of the raw materials. Finally four material variants were produced. The various configurations are given in table I.

Table I. Composition of material variants.

Material	Composition [vol.%]					
	Resin		Activated carbon		Carbon fibers	Wood fibers
	Urea-formaldehyde + additives	Phenolic	Type A	Type B		
MDF	15	-	-	-	-	85
ACBC-C2	-	30	-	70	-	-
ACBC-E2	-	18	-	69	13	-
ACBC-E7	-	18	69	-	13	-

The manufacture of biomorphic SiSiC ceramics is a three-stage process, divided into following practical parts. The process steps, which based on each other, are shown schematically in figure 1.

1. Green-body manufacture

The powdery ACBC compounds were transferred into form-stable and near net shape preform structures by using warm pressing procedure. In present work, all three ACBC green bodies were pressed in a planar steel die (reference samples) and additionally in a bowl shaped aluminum die (curved samples), at pressures up to 30 MPa and temperatures up to 200 °C.

The warm pressing process program, which regulates temperature curve and pressurization, was standardized. During filling the die a homogenous distribution of the compounds is essential to ensure a uniform microstructure and dimensional stability of preform.

Commercially available medium density fiberboards represent the plane green body preform of the material variant MDF. Due to the fact, first processing step of green body manufacture is omitted for material type MDF

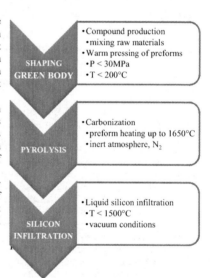

Figure 1. Schematic LSI-process depiction.

2. Pyrolysis / carbon preform manufacture

A full conversion of green body to porous carbon preform could be realized with the pyrolysis process step. Thereby, all specimens were heated up to 1650 °C in a flowing nitrogen atmosphere. This inert atmosphere ensured to remove all fission products and the complete conversion of polymer content to carbon. A furnace (type: OPUS 900, FCT Anlagenbau, Sonneberg, Germany) equipped with a carbon retort was used. During this process, mass loss arise with a corresponding volumetric shrinkage, which is critical regarding to near net shape manufacturability of complex formed structural components. All material variants were pyrolysed under the same conditions.

3. Siliconization

Finally, the highly porous carbon preforms were infiltrated with liquid silicon at a temperature of more than 1450 °C. This procedure is pressureless and passes by capillary forces. Every sample was infiltrated with sufficient silicon to ensure that the pore channel system is filled completely. The flowing silicon reacts with the carbon capillary walls and forms silicon carbide. After infiltration a biomorphic SiSiC structure with monolith character exists.

Specimen manufacture

Planar reference plates with dimension 300x300x10 mm³ were produced of each material variant. Green body processing parameters were always the same by manufacturing the ACBC material reference plates. Processing type MDF starts with drying the MDF plates and following pyrolysis step. At the end of LSI process chain, successively parts of the planar plates have been conserved for investigations. All specimens for determination of physical properties were taken from these planar reference plates. Additionally, curved specimens (100x100x4 mm³, curvature radius R = 97 mm) of ACBC material types were manufactured by warm pressing. Finally, resulting curved shaped preforms offered a wide spectrum of compaction degrees in occurring material modifications which deserved closer inspection.

Microstructure characterization

The microstructures of the materials were investigated after every processing step and thereby in three different microstructure configurations. Density and porosity values by samples of these configurations, could be determined by Archimedes method (according to DIN EN 993-1), pycnometer measuring (type: Accupyc 1330, Micromeritics Instrument Corporation, Norcross, Georgia, USA, acc. to DIN EN 1183), mercury intrusion porosimetry (type: Pascal 240, CE Instruments, Hindley Green, United Kingdom, acc. to DIN EN 66133) and CT investigation.

The determination of open and closed porosity in carbon preforms is an essential parameter for the last process step siliconization. For a sustainable analysis of green body and carbon preform pore channel system, Micro Computer Tomography scans of 12 samples have been performed using a high resolution µCT-System consisting of a microfocus X-ray tube with a maximum accelerating voltage of 180 kV and a 12-bit flat panel detector with an active area of 2300x2300 pixels at 50 microns per pixel. The CT specimen size was 4x4x10 mm³. The CT scans have been performed with X-ray parameters of 80 kV/200 µA and at an exposure time 3000 ms. Pixel size of 1.16 µm was attained at this configuration. The acquired X-ray images were reconstructed with a special reconstruction algorithm known as Filtered Back Projection. Each scanned volume was divided into 4 ROIs (Regions of Interest). Consequently, 48 reconstructed data sets had to be evaluated in total. Some image processing techniques such as median filtering were applied to the CT volume images in order to reduce the noise. This step contributed to an increase in the reliability of the image calibration based on gray scale methods. The purpose of the so called calibration or segmentation is to define both material and background allowing for the quantitative analysis of the CT data by evaluating features such as open and closed porosity, inclusions, etc. Due to applying mercury intrusion porosimetry, it was possible to make statements about pore size and pore size distribution of 6 investigated CT-specimens.

Moreover, the use of scanning electron microscope (type: Zeiss Ultra Plus, Carl Zeiss NTS GmbH, Oberkochen, Germany) was served for the detection of 2-dimensional microstructure composition after siliconization. Silicon carbide, residual silicon and carbon level were determined. In this context and to evaluate different compaction degrees regarding the silicon carbide conversion, polished and cross sections specimens were prepared out of the curved and planar SiSiC structures.

Shrinkage behavior
    A mass- and geometry shrinkage of the structure occurs during pyrolysis because conversion of polymer to carbon. For the investigated materials in this work, the polymer binder consists of phenolic resin (material type ACBC) as well as urea-formaldehyde resin (material type MDF). The higher the mass- and volume polymer percentage in preforms, the greater the conversion rate of structure after pyrolysis. To reduce the shrinkage, the materials ACBC E7 and ACBC E2 were spiked with carbon fibers. In this case, anisotropic shrinkage was implied. The shrinkage behavior of all materials during pyrolysis was investigated and compared. For this, cylindrical and plate-shaped green body specimens with different sizes were pyrolysed and statistically evaluated.

Mechanical Testing
    Flexural strength of SiSiC specimens was determined by four-point-bending test at room temperature on a universal testing machine (type: Zwick Roell 1475, according to DIN EN 843-1). One strain gauge on each test sample pull side was used to determine the Young's modulus. The specimens were 50x4x3 mm³ in length, width and height respectively. A loading rate of 0.5 mm min^{-1} was employed. Fracture surface images of each material were obtained by scanning electron microscopy to reveal the nature and origin of failure under these testing conditions. The Vickers hardness was measured according to DIN EN 6507-1 for the materials under the load of 9.81 Newton (HV1). A correlation between obtained mechanical property values and residual silicon and carbon presence could be made. Furthermore, thermal expansion rates for the temperature range between 20 and 1400 °C of investigated SiSiC ceramics were determined by dilatometer measurements (type: DIL 402 C, Netzsch Gerätebau GmbH, Selb, Bayern, Germany, acc. to DIN EN 1159-1).

RESULTS & DISCUSSION

Composite microstructures
    Starting with analyzing the planar reference plates, first microstructure information about different material configurations could be collected. Raw density and open porosity measured by Archimedes method and skeleton density measured via heliumpycnometer of specimens cut out from the reference plates were determined. The results are summarized in table II. In context of using the same processing parameters for all plates, the irregular proportionality between decreasing and increasing values of density suggested that differences in carbon modification and pore size distribution must be present. Furthermore, every carbon based material showed different compaction characteristics during warm pressing process. It was more difficult to compact material type E2 than type E7, due to the fact of voluminous carbon grains with size range 20 – 300 μm (activated carbon type B). Carbon type A is more fine-grained with grain sizes between 5 – 100 μm. By pressing type C2 without carbon fibers the effect was relativized, which is reflected in green body raw densities. In principle, the skeleton densities approximately were higher by the factor 2 than the raw densities in both high porous configurations green body and carbon preform. High open porosity values in a range of 40 – 60 % were the basis for silicon infiltration. The resulting SiSiC ceramics showed low open porosities and a good correlation between skeleton and raw density values. Finally, the densities range between 2.7 – 3.0 g/cm³, which could be obtained, is very close to pure silicon carbide with a density of approximately 3.2 g/cm³.

In relation to carbon preform 450 wt.% silicon was offered for every type of material during siliconization. The silicon uptake behavior of type MDF was in average 440 wt.%. Considering charcoal materials the silicon uptake was 270 wt.% for type E7, 290 wt.% for type E2 and 215 wt.% for type C2. Regarding net shape manufacture it would be sufficient to offer 300 wt.% silicon for the ACBC materials under these processing conditions. This adopted amount of silicon would avoid residual silicon on structure surface. Recent SiSiC microstructures of the reference samples are shown in Figure 2. The largest silicon carbide conversion ability was found in type ACBC E7 with 81 vol.%. Type MDF showed the highest residual silicon content and typical wood fiber derived silicon carbide grains. Cluster of residual carbon (size range 50 μm – 200 μm) in microstructure of type C2 indicated differences in pore size distribution compared to fiber reinforced materials. It may be emphasized, that not only the porosity influenced the silicon carbide conversion degree, but also the morphology of the pore channel system.

Table II. Material densities after LSI processing steps.

Material	Green body configuration			Carbon preform			SiSiC configuration		
	Density		Porosity	Density		Porosity	Density		Porosity
	Skeleton	Raw	Open	Skeleton	Raw	Open	Skeleton	Raw	Open
	[g/cm³]	[g/cm³]	[%]	[g/cm³]	[g/cm³]	[%]	[g/cm³]	[g/cm³]	[%]
MDF	1.41	-	-	1.38	0.51	62.70	2.75	2.77	0.12
ACBC-C2	1.50	0.86	39.24	1.45	0.90	36.26	2.65	2.71	0.14
ACBC-E2	1.62	0.80	45.16	1.51	0.76	48.40	2.95	2.96	0.14
ACBC-E7	1.63	0.86	46.13	1.56	0.82	46.06	3.00	3.00	0.13

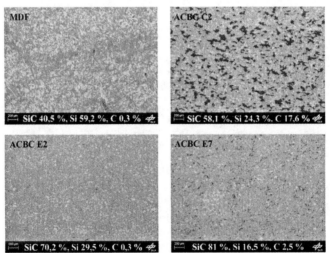

Figure 2. Typical microstructures of SiSiC materials. Black areas represents residual carbon, dark grey areas correspond with silicon carbide and light grey with residual silicon.

For each ACBC material and its compaction characteristics, different density gradients resulted in curved specimens after warm pressing. For evaluation the raw density and open porosity of every finite sample were measured via Archimedes method. By sampling 3 significant cross section areas a density gradient route was observed. The density gradients in cross section areas of curved specimens and associated specimen sampling description are shown in fig. 3.

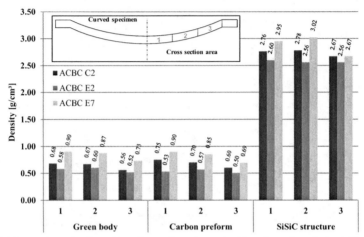

Figure 3. Measured raw density values of curved specimens in all configurations during LSI-route.

After warm pressing, the average density value of type ACBC E7 in green body configuration was nearly the same compared to the plane reference plate. Despite constant processing parameters the types E2 and C2 did not show corresponding density values of reference plates. One reason might be the large grain size of activated carbon type B and its worse compaction characteristics. Furthermore, inhomogeneous compound distribution during filling the curved aluminum die may be another reason for the density gradients in specimen structure.

To extract information about the pore channel system, 2- and 3-dimensional CT images were taken by curved specimens placed in cross section area 1 and 2 for green body and carbon preform configuration. Parts of pore channel system of the samples material C2 (a, b), E2 (c, d) and E7 (e, f) in carbon preform configuration are shown in fig. 4. It might to be reasonably assumed that the carbon fibers reduce the maximum material wall thicknesses. Fundamentally, no significant differences between the carbon fiber containing materials ACBC E2 and E7 and the fiber free material ACBC C2 could be identified. All materials showed a very complex formed channel system with completely connected pore channels and channel branches - in x, y and z orientation. The closed pores as well as foreign inclusions were in a range of 0.1 – 0.2 vol.% and therefore very small compared to total investigated volume. Thus, residual carbon in SiSiC configuration based almost exclusively on excess carbon in the walls, which could not be full converted. Open porosities determined by means of CT investigations are slightly smaller than values measured by Archimedes method which might be caused by measuring errors. Beyond this, CT investigations offered only a finite point of view of total volume pore distribution. The results of each investigated finite specimen area are summarized in table III.

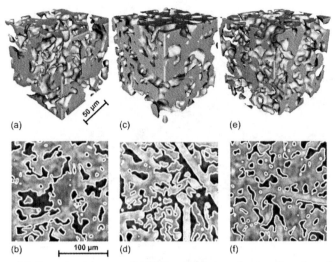

Figure 4. 3- and 2-dimensional images of ACBC materials in carbon configuration after pyrolysis. Pore channel system (colored in 3D) out of cross section area 1 from specimen C2 (a, b), E2 (c, d) and E7 (e, f).

Table III. Results of CT investigations obtained with corresponding CT evaluating software.

Configuration	Green body						Carbon preform					
Specimen	C2		E2		E7		C2		E2		E7	
cross section area	1	2	1	2	1	2	1	2	1	2	1	2
Investigated volume V [mm³]	0.8	0.8	0.8	0.8	0.8	0.8	0.8	0.8	0.8	0.8	0.8	0.8
Open porosity e' [%]	39.74	39.34	26.95	46.46	40.21	39.52	32.29	36.52	57.01	49.41	34.19	41.22
Closed porosity e'$_{closed}$ [%]	0.14	0.19	2.8	0.04	0.04	0.03	0.26	0.15	0.04	0.04	0.08	0.13
Foreign inclusions V$_{Fi}$ [%]	0.24	0.28	0.24	0.21	0.28	0.21	0.04	0.03	0.02	0.03	0.22	0.16
Material volume V$_M$ [mm³]	0.48	0.48	0.56	0.43	0.48	0.48	0.54	0.51	0.34	0.4	0.52	0.47

Pore size distribution of CT-specimens out of cross section area 1 in green body as well as carbon preform configuration was measured by mercury intrusion porosimetry (table IV). All materials showed a homogenous pore size distribution in green body and carbon preform configuration. Significant peaks of modal pore diameter have always been present (fig. 5). For type ACBC C2 and E2 the modal diameter was ~9 μm in green body (fig. 5 a, c) and between 7 – 10 μm in carbon preform (fig. 5 b, d). Type ACBC E7 exhibits smaller pore sizes with a modal pore diameter of ~5 μm in green body and ~4 μm in carbon preform (fig. 5 e, f). This trend has continued over measured pore surface and average pore diameter values (table IV). This fact is an indication for main influence of activated carbon grain size on pore size and pore size distribution. Comparing modal pore size in carbon stage with resulting SiSiC microstructure and density in cross section area 1 of material type E7 suggested that silicon infiltration and silicon carbide conversion for pore sizes of about ~4 μm is near optimum.

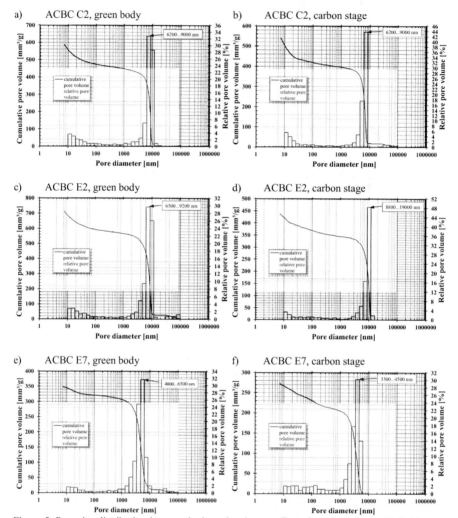

Figure 5. Pore size distribution in green body and carbon configuration of material ACBC C2 (a, b), ACBC E2 (c, d) and ACBC E7 (e, f). Highest peak represents the modal diameter range of each material type and configuration.

Table IV. Pore size and pore surface volume measured by mercury intrusion porosimetry in green body and carbon preform configuration. Measured specimens were cut from cross section area 1 from curved structures.

Material	Configuration	Total pore volume	Total pore surface area	Average pore diameter
		[mm³/g]	[m²/g]	[nm]
C2	Green body	585.46	31.65	73.99
E2	Green body	715.05	33.44	85.53
E7	Green body	348.08	7.02	198.43
C2	Carbon preform	540.53	37.93	57.01
E2	Carbon preform	438.13	18.69	93.79
E7	Carbon preform	272.98	9.65	113.13

In curved parts which were processed by warm pressing density gradients occurred in the carbon stage which had a considerable impact on subsequent silicon carbide yield and SiC conversion during silicon infiltration. This was confirmed by scanning electron microscope images of siliconized curved specimens. For analysis specimens were cut out of 3 significant cross section areas. First of all, every type of material showed residual silicon content which was higher by a factor of 2 compared to the reference plates. Reason for this was the high open porosity of the curved carbon preforms (see fig. 3). Additionally, the silicon carbide-, residual silicon- and residual carbon contents were different within a material cross section. As example the density gradient in ACBC E7 is shown in fig. 6. Cross section 1 (was high dense in carbon preform, 0.90 g/cm³) showed nearly the same SiC conversion degree as the middle section 2 (carbon preform density = 0.85 g/cm³). The only difference between these cross section areas is the residual carbon content which is higher in section 1 (Fig. 6, E7-1 / E7-2). In total, the SiC conversion degree of ~70 vol.% was smaller than the conversion degree in reference plate ACBC E7 (81 vol.%), although density and open porosity in carbon preform of reference plate and curved specimen were in good agreement. Thus, pore size and pore size distribution are very important for silicon carbide formation during infiltration of highly porous carbon. A homogenous SiC conversion degree in cross section of curved structures required an optimization of warm pressing technique.

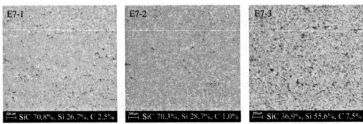

Figure 6. Microstructure configurations in cross section areas (1 – 3) of curved specimen ACBC E7.

Shrinkage behavior
All materials showed differences in shrinkage behavior, depending on material composition. The estimated percentage values of geometry- and mass shrinkage are compared and listed in fig. 7. The fiber doped materials, ACBC E2 and ACBC E7, were found to have the slightest shrinkage and showed their potential for net-shape manufacturing of SiSiC structures. The shrinkage of material type

C2 was reduced by ~50 % compared to wood based material MDF. Because of its high shrinkage values and inflexibility in processing complex structures, material type MDF is not a candidate for net-shape manufacture under these processing conditions.

On macroscopic scale, there were no visible cracks on curved and plane sample surfaces of ACBC materials. This good shrinkage behavior indicates that complex shaped components can be produced using this composition.

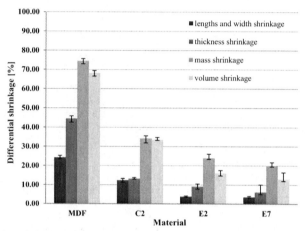

Figure 7. Shrinkage behavior during pyrolysis of wood fiber based material (MDF) and activated carbon based materials (C2, E2 and E7).

Physical properties

Table V summarizes the physical properties of the four investigated biomorphic SiSiC ceramics. Flexural strength, Young's modulus and hardness values increase with growing silicon carbide content. It was found that certain residual carbon content could be eradicated with simultaneous presence of high specific silicon carbide content. Images of typical fracture surfaces after four-point-bending test of each type of material are faced in fig. 8. These crack surfaces are a result of medium to high energy fracture behavior. The mean linear thermal expansion value differed only marginally for temperature range 20 – 1400 °C.

Table V. Physical properties of biomorphic inspired SiSiC materials.

Material	Physical properties				
	Density	Flexural strength	Young's modulus	Vickers hardness	Mean linear thermal expansion
	[g/cm³]	[MPa]	[GPa]	HV1 [GPa]	[10^{-6} K^{-1}] (20-1400C°)
MDF	2.77 ± 0.01	139 ± 25	243 ± 23	2014 ± 364	4.48 ± 0.05
ACBC-C2	2.71 ± 0.01	156 ± 24	232 ± 58	2257 ± 348	4.45 ± 0.05
ACBC-E2	2.96 ± 0.01	190 ± 20	287 ± 34	2395 ± 222	4.59 ± 0.05
ACBC-E7	3.00 ± 0.01	201 ± 21	325 ± 25	2451 ± 321	4.52 ± 0.05

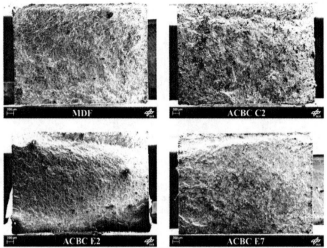

Figure 8. Typical crack surfaces of four-point-bending test specimens.

CONCLUSION

Three charcoal and one wood fiber based biomorphic SiSiC ceramics with different microstructures were successfully produced via LSI-technique. The critical shrinkage behavior of wood fiber based materials during processing could be reduced by over 50 % by the use of charcoal based compounds. It could be shown that compounds based on activated carbon are promising candidates for structures with high silicon carbide content and complex forms.

First experiments of processing curved structures via warm pressing with ACBC compounds were conducted. Thereby, pore channel system in green body and carbon preform configuration could be investigated more closely. Furthermore it was possible to demonstrate the high impact of density gradient, transferred by warm pressing of curved structures, on pore size and pore size distribution and depending silicon carbide conversion. The focus for future work will be the processing of complex structures without density gradients and high silicon carbide contents.

Physical properties of investigated materials were determined. Silicon carbide content in microstructure is the most important factor for mechanical properties of SiSiC ceramics. However, the ratio between residual silicon and carbon is not as relevant in conjunction with significant silicon carbide presence.

ACKNOWLEDGEMENT

This research was supported by German Aerospace Center, Institute of Structures and Design, Stuttgart, Germany. The authors gratefully acknowledge Christoph Hillebrand and Jürgen Horvath (University of Bremen, Germany) for carrying out hardness measurements and Gudrun Steinhilber for conducting mercury intrusion porosimetry measurements.

REFERENCES
[1]S. Kumar, A. Kumar, A. Shukla, A.K. Gupta, R. Devi, Capillary infiltration studies of liquids into 3D-stitched C–C preforms: Part A: Internal pore characterization by solvent infiltration, mercury

porosimetry, and permeability studies, *Journal of the European Ceramic Society 29*, 2643 - 2650, 2009.

[2]S. Kumar, A. Kumar, A. Shukla, A.K. Gupta, R. Devi, Capillary infiltration studies of liquids into 3D-stitched C–C preforms: Part B: Kinetics of silicon infiltration, *Journal of the European Ceramic Society 29*, 2651 - 2657, 2009.

[3]P. Greil, T. Lifka, A. Kaindl, Biomorphic Cellular Silicon Carbide Ceramics from Wood: I. Processing and Microstructure, *Journal of the European Ceramic Society 18*, 1961, 1998.

[4]D.-W. Shin, S.S Park, Y.-H. Choa, K. Niihara, Silicon / silicon carbide composites Fabricated byinfiltration of a silicon melt into charcoal, *Journal of the American Ceramic Society 82*, 3251–3254, 1999.

[5]T. Xue, Z.J.W. Wang, Preparation of porous SiC ceramics from waste cotton linter by reactive liquid Si infiltration technique, *Materials Science and Engineering: A, 527*, 7294 - 7298, 2010.

[6]M. Singh, D.R. Behrendt, Microstructure and mechanical properties of reaction-formed silicon carbide (RFSC) ceramics, *Material Science and Engineering*, 183-187, 1994.

[7]T.M. Lillo, D.W. Baily, D.A. Laughton, H.S. Chu, W.M. Harrison, Development of a pressureless sintered silicon carbide monolith and special-shaped silicon carbide whisker reinforced silicon carbide matrix composite for lightweight armor application, *Ceramic engineering and science proceedings*, 359-364, 2003.

[8]Y. Wang, S. Tan, D. Jiang, The effect of porous carbon preform and the infiltration process on the properties of reaction-formed SiC, *Carbon 42*, 1833 - 1839, 2004.

[9]L. Hozer, J.R. Lee, Y.M. Chiang, Reaction-infiltrated, net-shape SiC composites, *Department of Materials Science and Engineering*, 131-143, 1995.

[10]P. Greil, Near net shape manufacturing of ceramics, *Materials Chemistry and Physics 61*, 64-68, 1999.

[11]J. Qian, Z. Jin, Preparation and characterization of porous, biomorphic SiC ceramic with hybrid pore structure, *Journal of the European Ceramic Society 26*, 1311–1316, 2006.

[12]Y. Ikeda, M. Yasutoshi, M. Mizuno, M. Mukaida, M. Neo, T. Nakamura, 3 Dimensional CT analyses of bone formation in porous ceramic biomaterials, *Ceramic engineering and science proceedings*, 185-190, 2003.

BIOFLUID FLOW SIMULATION OF TISSUE ENGINEERING SCAFFOLDS WITH DENDRITE STRUCTURES

Satoko Tasaki, Chiaki Maeda and Soshu Kirihara
Joining and Welding Research Institute, Osaka University
11-1, Mihogaoka, Ibaraki, Osaka

ABSTRACT

Many materials and structures have been manufactured for bone tissue engineering via rapid prototyping. Computer aided design and manufacturing process can efficiently create products. However, bioscaffold evaluation takes a significant amount of time. In this investigation, biofluid flow in various dendrite bioscaffolds with the same porosity was simulated. The simulated models were four-, six-, eight-, and twelve-coordinate structures. The aspect ratio of the rods in the structures was adjusted to control porosity. Flow velocity in the spatial grids in the models was calculated through the finite element method.

INTRODUCTION

Tissue scaffolds are required to repair bone defects resulting from illnesses. To encourage osteoconductivity and tissue regeneration, it is important that prosthetics mimic bone porosity and optimized flow behavior.

Various rapid prototyping techniques for scaffolds have been investigated. For example, Williams, et al. [1] created polycaprolactone scaffolds with variable pore size and porosity from 63% to 79% by using selective laser sintering. Seitz, et al. [2] fabricated 3D printed scaffolds using modified hydroxyapatite. Simon, et al. [3] used direct ink writing for periodic array scaffolds. Previously, we reported bioceramic implants for dental crowns and hydroxyapatite scaffolds fabricated by stereolithography [4, 5] Considerable research has been conducted on tissue engineering for computer aided design and manufacturing (CAD/CAM).

The scaffold structure requires suitable porosity and pore size to foster tissue regeneration in the human body [6]. However, evaluating osteogenesis requires long-term clinical experiments. As an alternative to clinical experiments, in this investigation, we fabricated hydroxyapatite dendrite scaffolds, simulated biofluid flow behavior, and considered the effects of four different types of architecture with the same porosity.

EXPERIMENTAL

Four-, six-, eight-, and twelve-coordinate (= dendrite) structures were designed with CAD software (Materialize, Magics Ver. 14). Figure 1 illustrates the four dendrite scaffolds. Aspect ratios (rod length to diameter ratios) were adjusted in the scaffold models to control porosity in the range of

50%–90%. The porosity of all skeletal structures was 75%, which is the same as the porosity of a human bone. These scaffolds have 100% interconnected pores.

Computer graphic models of the structures were automatically converted into a numerical data format and sliced into file sets of cross-sectional planes with uniform thickness. These operational data sets were transferred to the stereolithography apparatus (D-MEC, SCS-300P) of the CAM equipment. Photosensitive acrylic resin (JSR Corporation, KC-1159-2) and 45 vol. % hydroxyapatite powder (grain size avg: 5–20 μm, Taihei Chemical Industrial, Hap-200) were used for the slurry material in this investigation. These materials were mixed using a planetary centrifugal mixer (Thinky, AR-250) for 5 min. The slurry was spread on a flat stage and then smoothed. An ultraviolet laser beam (λ = 355 nm) was scanned over the deposited layer to create cross-sectional planes. Through layer-by-layer processes, solid components were fabricated. The fabricated precursors were dewaxed at 600°C for 2 h at a heating rate of 1.0°C/min [7] and sintered at 1250°C for 2 h at a heating rate of 5.0°C/min in air [8]. The hydroxyapatite porous body form and microstructure were observed using an optical microscope (Keyence, VH-Z 100) and a scanning electron microscope (SEM: JEOL, JSM 6060), respectively. The sintered densities were measured by the Archimedes method [9].

Fluid circulation in the various dendrite scaffolds was visualized with the fluid dynamic solver (Cybernet System, ANSYS ver. 13). Flow velocity in the spatial grids in the scaffold models was calculated through the finite element method (FEM). The following values of these parameters were used in this simulation [10, 11]. The fluid phase was represented as an incompressible Newtonian fluid with a viscosity of 1.45×10^{-3} Pa·s. The inlet velocity applied to the scaffolds was constant at 0.235 mm/s, and the pressure was zero at the outlet. No-slip surface conditions were assumed.

RESULTS AND DISCUSSION

The hydroxyapatite ceramic scaffolds form and microstructure are shown in Figs. 2 (a) and (b), respectively. Micrometer-order ceramic lattices with periodic structures were successfully fabricated using CAD/CAM processes to realize a diamond-type (four-coordinate) structure, as shown in Fig. 1(a). The porosity of the scaffold form was approximately 75%. The relative density of the hydroxyapatite ceramics was 98%.

Figure 3 shows the streamline behavior in the dendrite scaffolds. Figure 3 (a) shows a four-coordinate scaffold structure and indicates that inordinate flow at a low velocity was obtained. This also indicates simulated biofluid flow to the whole structure, which is expected to provide active tissue regeneration. The fluid velocity of the six-coordinate scaffolds is the highest at above 1.0 mm/s (Fig. 3 (b)). There are no blockades from the inlet to the outlet, and the flow becomes linearly stable. The high fluid velocity area (above 1.0 mm/s) in the scaffold is subjected to shear stress which can assume the difficulty cell attachment on the scaffolds surface [12]. The eight-coordinate structure (Fig. 3 (c)) simulated random fluid flow and high velocity in some parts of the structure. This is because it has more obstacles than the four- and six-coordinate scaffolds and the number of coordinate rods is

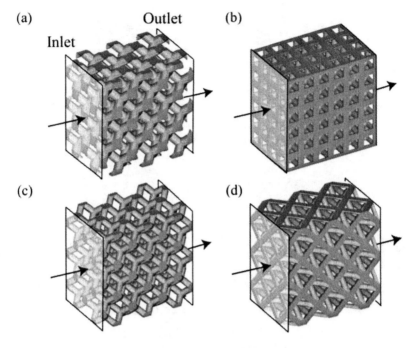

Figure 1. Dendrite scaffolds models.
(a) Four coordinate. (b) Six coordinate. (c) Eight coordinate. (d) Twelve coordinate.

increased. The twelve-coordinate scaffold structure (Fig. 3 (d)) has a moderate velocity and flow behavior, similar to the four-coordinate structure. However, there are no active flows in some areas because of many rods in the architecture.

CONCLUSION

We successfully fabricated dendrite scaffolds and visualized fluid flow in them. Various structures with the same porosity exhibited different flow behavior and velocity via fluid simulation. The results suggest that it is reasonable to assume that the four- and twelve-coordinate dendrites are optimal. The scaffolds with exactly aligned pores have a higher fluid velocity, and hence, the surface areas become susceptible to shear stress. Structural modification may help in effective tissue regeneration.

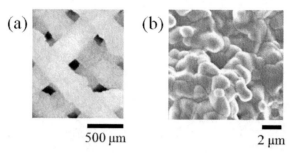

500 μm   2 μm

Figure 2. Fabricated hydroxyapatite scaffolds with dendrite structure.
(a) Scaffold form. (b) Microstructure.

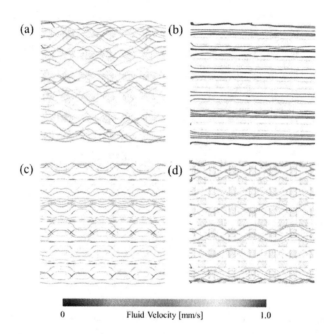

0   Fluid Velocity [mm/s]   1.0

Figure 3. Fluid flow behaviors in the dendrite scaffolds visualized using the finite element method.
(a) Four coordinate. (b) Six coordinate. (c) Eight coordinate. (d) Twelve coordinate.

ACKNOWLEDGEMENT

This study was supported by Priority Assistance for the Formation of Worldwide Renowned Centers of Research–The Global COE Program (Project: Center of Excellence for Advanced Structural and Functional Materials Design) from the Ministry of Education, Culture, Sports, Science and Technology (MEXT), Japan.

REFERENCES

[1] J. M. Williams, A. Adewunmi, R. M. Schek, C. L. Flanagan, P. H. Krebsbach, S. E. Feinberg, S. J. Hollister, and S. Das, Bone tissue engineering using polycaprolactone scaffolds fabricated via selective laser sintering, Biomaterials, **26**, 4817–4827 (2005)

[2] H. Seitz, W. Rieder, S. Irsen, B. Leukers, and C. Tille, Three-dimensional printing of porous ceramic scaffolds for bone tissue engineering, J. Biomed. Mater. Res. B: Appl. Biomater., **74** [2] 782-788 (2005)

[3] J. L. Simon, S. Michna, J. A. Lewis, E. D. Rekow, V. P. Thompson, J. E. Smay, A. Yampolsky, J. R. Parsons, and J. L. Ricci, In vivo bone response to 3D periodic hydroxyapatite scaffolds assembled by direct ink writing, J. Biomed. Mater. Res. A, **83A** [3] 747-58 (2007)

[4] C. Maeda, S. Tasaki, and S. Kirihara, Accurate Fabrication of Hydroxyapatite Bone Models with Porous Scaffold Structures by Using Stereolithography, IOP Conf. Ser.: Mater. Sci. Eng., 18 [072017] 1-5 (2011)

[5] S. Tasaki, S. Kirihara, and T. Sohmura, Fabrication of Ceramic Dental Crowns by using Stereolithography and Powder Sintering Process, Advanced Processing and Manufacturing Technologies for Structural and Multifunctional Materials V: Ceramic Engineering and Science Proceedings, **32** [8] 141-146 (2011)

[6] V. Karageorgiou, and D. Kaplan, Porosity of 3D biomaterial scaffolds and osteogenesis, Biomaterials, **26** [27] 5474-5491 (2005)

[7] M. Suwa, S. Kirihara, T. Sohmura, Fabrication of Alumina Dental crowns Using Stereolithography, Ceramic Transactions, 219, 331-336 (2009)

[8] B. Chen, T. Zhang, J. Zhang, Q. Lin, D. Jiang, Microstructure and mechanical properties of hydroxyapatite obtained by gel-casting process, Ceram. Int., 34, 359-364 (2008)

[9] W. D. Kingery, H. K. Brown, D. R. Uhlman, Introduction to ceramics, Second ed., Wiley, New York, 530-531, (1976)

[10] A. L. Olivares, E. Marsal, J. A. Planell, and D. Lacroix, Finite element study of scaffold architecture design and culture conditions for tissue engineering, Biomaterials, **30** 6142–6149 (2009)

[11] A. J. F. Stops, K. B. Heraty, M. Browne, F. J. O'Brien, and P. E. McHugh, A prediction of cell differentiation and proliferation within a collagen–glycosaminoglycan scaffold subjected to mechanical strain and persuasive fluid flow, Journal of Biomechanics, 43 618-626 (2010)

[12] C. Jungreuthmayer, M. J. Jaasma, A. A. Al-Munajjed, J. Zanghellini, D.J. Kelly, and F. J. O'Brien, Deformation simulation of cells seeded on a collagen-GAG scaffold in a flow perfusion bioreactor using a sequential 3D CFD-elastostatics model, Med. Eng. Phys., 31 420-427 (2009)

Porous Ceramics

MULTIFUNCTIONAL CARBON BONDED FILTERS FOR METAL MELT FILTRATION

Christos G. Aneziris[*], Marcus Emmel*, Anja Stolle**

*Institute of Ceramic, Glass and Construction Materials, Technical University of Freiberg, Germany
** Institute of Metal Casting Technology

ABSTRACT

New carbon bonded filters with alumina based "active coatings" for higher filtration efficiency of alumina based inclusions as well as nano-engineered filters with nano-scaled additives are explored with the aid of impingent tests and are evaluated according to their cold crushing strengths at room temperature. The combination of carbon nanotubes and alumina nanosheets additives leads to in situ formation of Al_3CON. Both the nano-scaled additives as well as the extra alumina coating lead to improved mechanical performance of the carbon bonded filters and open the horizon for filter macro structures with higher filtration capacities in means of bigger dimensions.

INTRODUCTION

Small particles so called inclusions in the range of some microns can stop a train. Based on a failure analysis report of the Federal Institute for Material Research in Berlin clustered alumina inclusions have led to the failure of the wheel and axle set of the ICE-3 train of the German Railway in 2008.[1] There exists an increasing pressure on the metal-making and metal-using industry to remove solid and liquid inclusions such as deoxidation products (oxides), sulfides, nitrides carbides etc. and thereby improve metal cleanliness. It is well known that size, type and distribution of non-metallic inclusions in metal decrease dramatically the mechanical properties and especially the fracture toughness, the tensile strength, the ductility as well as the fatigue of the cast products resulting to excessive casting repairs or rejected castings.[2] In case of the oxide inclusions in steel melts Wasai et al. assigned the dendritic, maple-like and polygonal inclusions to the group of the primary inclusions generated directly after giving aluminum in the metal melt.[3] In contrast, the network-like, coral-like and spherical inclusions, which are composed of alumina, hercynite and wüstite, are classified as secondary inclusions. The secondary inclusions are formed due to the lower solubility of oxygen in the melt as a function of the temperature above the liquidus temperature of the melt. Regarding the secondary alumina-inclusions, especially the α-, γ- and δ-modification are more frequently detected. Below the liquidus temperature tertiary and quartenary inclusions are generated, that present the highest impact on fractures toughness of steel casts according to Ovtchinnikov.[4]

In the last four decades ceramic foam filters are successfully applied in iron and aluminium foundries for metal products with superior properties. Two basic functions fulfil nowadays filter systems a) remove impurities in millimetre, micron and submicron area and b) promote none turbulent melt filling of mould reducing possibility of reoxidation and erosion. Concerning the thermo mechanical requirements of filter materials and filter structures the thermal shock resistance, the refractoriness under load as well as sufficient mechanical strength at temperatures in the range of 1500°C for iron and 1650°C for steel are needed. In case of large castings bigger casting surfaces as well as longer casting times with improved creep resistance of the filter components are required. For the steel melt filtration filters based on partially stabilised zirconia are established since the last two decades. In order to reduce shrinkage defects as they appear in oxide filter systems, improve thermal shock resistance, increase the creep resistance and also achieve higher filtration capacities by reducing total filter costs, the carbon bonded ceramic technology has been successfully applied for steel filtration. Comparing a carbon bonded alumina ceramic filter (CBC) and a ceramic bonded zirconia filter, in spite the fact that the same porous polymer macrostructure of 200mm in diameter have been used, with the CBC filter 1500kg steel melt have been filtered at 1620°C against 1000 kg by the zirconia filter, after filtration.[5] These results show that a higher specific surface and a higher creep resistance (no deformation during casting) led to improved filtration capacities.

Janke et al. and Hammerschmid give an overview of the filtration mechanisms, filtration efficiencies, filter materials and filter structures for steel melts.[6,7] According to Dávila-Maldonado et al. the filtration efficiency of non-metallic inclusions in ferrous melts with the aid of ceramic foam filters is less than 75 % in the particle size range from 1 up to 100μm.[8]

The main production technology of ceramic filters is based on the patent of Schwartzwalder[9], whereby a polymer foam is impregnated in a ceramic slurry (this first coating contributes as an adhesive porous layer for further coating processes), the ceramic slurry is squeezed out of the functional pores and then spray coatings are applied in order to

a) eliminate defects out of the squeezing process and

b) reach critical wall thickness (strut thickness) of foam ceramic filter for acceptable thermomechanical and chemical strengths at elevated temperatures by fulfilling also the requirements of a special heat balance (avoiding freezing of the metal melt in the filter). For a foam macrostructure an optimum is achieved by a wall thickness value of approximately 0.3 mm. The influences of the viscosity of the impregnation slurry on the quality of the foams were investigated by Zhu et. al. whereby higher viscosities of the impregnation slurry led to a better covering of the struts of the polyurethane foam.[10] Saggio-Woyansky et. al. gave an overview about several methods to influence the surface of the polyurethane foam to result in a thicker layer of slurry after impregnation.[11] The quality of the ceramic filter is adjusted by the spraying coatings. For this reason the rheology of the spray slurry and the adjustment of the spray system have been investigated in Hasterok et al with the aid of computer tomography.[12]

The Collaborative Research Center (CRC) 920 "Multifunctional Filters for the Metal Melt filtration – a Contribution to Zero Defect Materials" granted by the German Research Foundation in 2011 at the University of Freiberg aims to improve the mechanical properties of metallic materials for security and light-weight constructions particular for applications in mechanical engineering and vehicle construction with the aid of surface functionalized filter materials based on so called "active" as well as "reactive" coatings. In case of the "active" coatings, the same chemistry as the chemistry of the primary or secondary inclusions that have to be removed are generated on carbon bonded filters. On the other hand "reactive" coatings react with the dissolved in the melt gas (for instance oxygen in steel melts) and create inclusions above the liquidus temperature of the melt that are deposit on the filter. With this approach less tertiary and quartenary inclusions are generated below the liquidus temperature. In the past Antsiferov and Porozova have presented "active" coatings for improved properties of cast iron and aluminum products and Aneziris et al have investigated so called "repairing coatings" based on sugar solutions for more flexible structures with less friability at room temperature.[13, 5]

In terms of this contribution preliminary results of the work of the CRC will be demonstrated. Firstly a new resin free carbon bonded filter will be introduced. Due to ecological and economical demands pitch based binders with a high Benzo[a]pyrane amount are not allowed any more for European filter manufacturers. The first topic deals with the application of a synthetic, environmental-more-friendly pitch in the impregnation and spray slurry of alumina carbon bonded filters. The synthetic pitch contributes as binder as well as a carbon source. Another topic deals with active coatings based on alumina generated on the new carbon bonded filters. Due to their high creep resistance these carbon bonded substrates are sufficient for the high thermo mechanical requirements of steel melt filtrations at temperatures between 1620 up to 1680°C, and due to the active coating a high filtration efficiency above 90% for the primary and secondary inclusions in the range of 1 up to 100μm is aimed. A last contribution deals with the application of nanometer additions based on carbon nanotubes (CNTs) and alumina nanosheets (NS) in the impregnation as well as in the spray slurry. With the aid of electron backscatter diffraction analyses (EBSD) on fracture surfaces of the carbon bonded samples, Al_3CON was identified on the nanosheet shapes already at 1.000°C coking temperature. The Al_3CON new phase based on the reaction between alumina nanosheets and CNTs offers a chemical interconnecting phase for the carbon as well as for the oxide alumina filler. Due to the nano-engineered foam filters macro-structures greater than 200mm in diameter for filtration of large steel castings can be aimed in the future.

EXPERIMENTAL

The raw materials, used for the preparation of the new alumina carbon bonded composition with the new synthetic pitch based on Carbores® P as a binder as well as carbon source, were calcined alumina (99.80 wt.-% Al_2O_3, ≤ 0.10 wt.-% Na_2O) with a d_{90}≤3.0μm (Martinswerk, Germany), Carbores® P (Rütgers, Germany) 0-20 μm, fine natural graphite grade AF 96/97 (99.8 wt.-% < 40 μm) with 96.7 wt.-% carbon content (Graphit Kropfmühl, Germany) and carbon black powder with a mesh plus of 30ppm<44μm (Lehmann & Voss & Co., Germany).

Table I. Compositions of "Carbores Filter" and the alumina based active coating.

	Carbores Filter	Active coating
	(wt.%)	(wt.%)
Impregnation slurry	82 solids	
Spray coating	74 solids	74 solids
Al_2O_3 Martoxid MR 70	66	100
Carbores® P	20	
Graphite AF 96/97	8	
Carbon black	6	
Ligninsulfonate* T11B	1.5	1.5
PPG P400*	-	0.8
Castament VP 95L*	0.3	-
Contraspum K 1012*	0.1	0.1
*: related to the solids		

Table II. Technical data of nanoscaled additives.

Nanoscaled additives	Producer	Purity (wt.-%)	Average particle size (nm)	Specific surface area (m^2/g)
Carbon Nanotubes (CNTs)	Timesnano (China)	> 95.0	-	> 200
Alumina Nano sheets (NS)	Sawyer (USA)	95.0 – 99.8	10 – 250	9 – 40

For the filter production templates based on polyurethane foam are used (pore size distribution in the range of 0.5 up to 5 mm) with a porosity of 10 pores per inch (ppi), a height of 22 mm and a diameter of 50 mm. The impregnation slurry consisting of 82% solids was produced in a ToniMIX high shear mixer (ToniTechnik, Germany). In the first step the powder components of the mixture (Table I) were dry mixed for 5 minutes. After this the liquid components based on deionised water, Lignisulfonate T11B (Otto Dille, Germany), Castament VP 95L (Bash, Germany) and Contraspum K 1012 (Zchsimmer & Schwarz, Germany) were added. Stepwise addition of the additional water led to a plastic mass, which was kneaded for another 5 minutes. While going on mixing the slurry was formed with adding the rest of the water. For production purposes the carbon bonded spray slurry consisting of 74% solids was homogenised with a VISCO JET stirrer system (Inotec, Germany). The same composition as the impregnation slurry was used, but a higher amount of deionised water was needed. The pure alumina based slurry for the active spray coating is also included in Table I and was also produced in the stirrer system. In this case a wetting additive based on PPG (Sigma Aldrich, Germany) was additionally added.

The slurries were characterised using a rheometer (RS 150, Haake, Germany) with a coaxial cylindrical measurement system (DIN 53019) and Z40 DIN system. The viscosity was determined in dependence of the shear rate and the yield stress was measured. The viscosities of the impregnation and spraying slurries were determined with a shear rate of 200 1/s and 1000 1/s, respectively. The measurement of the viscosity starts after a relaxing time of 20 s when the shear rate increases in 300 s to 1000 1/s. After a holding time of 60 s at 1000 1/s it decreases within 300 s to zero. Yield stress was measured after a slurry relaxing time of 60 s. Shear stress was increased by 1.85 Pa/s and the resulting deformation was measured.

All foams were impregnated with the impregnation slurry to form a first layer on the polyurethane skeleton. The second layer was applied through spraying onto both sides of the impregnated foam with defined conditions. The spraying was performed using a SATAjet B spraying gun and with a SATAjet B 1.0 E nozzle type (SATA, Germany). Between foam and spraying gun a constant distance of 27 cm as well as a constant pressure of the compressed air of 3 bar was maintained. The viscosity of the spraying slurries was varied through changes in the solid content. The slurry flow rate was adjusted by the aid of a regulating screw. Feedback was received as the mass of the slurry of a 30 s spraying was measured. A flow rate of 20 g/30s for a spraying time of 6 s was set.

In case of the nano-engineered filter compositions three filter versions were produced, a) impregnation slurry as "Carbores filter" with a spray slurry with 0.15 wt.% CNTs and 0.15 wt.% NS related to the solids, b) impregnation as well as spray slurry as "Carbores filter" with 0.15 wt.% CNTs and 0.15 wt.% NS related to the solids and c) impregnation as well as spray slurry as "Carbores filter" with 0.3 wt.% CNTs related to the solids. The nano-scaled additives were added in several steps during mixing. The technical data of the nano-scaled additives are listed in Table II.

After each impregnation as well as after each spraying step all samples were dried at 25°C for 12 h. Dried samples without any nano-scaled additives as well as without an active coating were cooked in a retort filled with coke breeze at 800°C for 180 min. The heating profile was 1K/min and at each 100K step 30min holding time was programmed. Taken under consideration the contribution of Roungos et al the carbon bonded filters with nanoscaled additives were coked at 1000°C under reduced conditions.[14] For the production of the carbon filters with the alumina based active coating, the carbores substrate filter was in advanced coked at 800°C, the active spray coating was applied at room temperature and then dried, and the composite filter was then coked at 1400°C for 300min also in a retort filled with coke breeze under reduced conditions. The coking behaviour of carbon bonded filters as well as of carbon bonded filters with an alumina based active coating have been monitored in a sintering dilatometer DIL 402 C (Netzsch Germany) in argon atmosphere. The carbon bonded filter was coked in advance and its thermal deformation was measured in the dilatometer. In case of the "active filter" the carbon bonded substrate was coked in advance and then the coking of the active coating took place in situ in the dilatometer.

After coking all carbon bonded filters have been evaluated due to a so called "small impingement" test of 5kg steel melt (42CrMo4 steel quality) free casting in air at 1670°C from approximately 50cm priming height, Figure 1. Normally for industrial oriented impingement tests 50kg melt for iron melts and approximately 100kg for steel melts are used. The small impingement test is a first indication according to the thermal shock performance of the new produced filters.

Figure 1. "Small impingement" test of carbon bonded filter with 20 wt% Carbores® P, 5kg steel melt 42CrMo4 at 1670°C.

The determination of the cold crushing strength (CCS) was carried out at the universal testing machine TT 2420 (TIRA GmbH, Germany), with a pressure cell of 20kN. The speed adopted was 3 mm/min, until a counteracting force of 5 N was reached. This was the switch-point, starting from which the speed was regulated to 1 N/mm²/s. When a loss of strength of 80% was reached, the measurement was terminated. The open porosity of the struts of the coked carbores filter was measured with the aid of a mercury porosimeter.

In order to detect the strut thickness distribution, the foams were analyzed with the aid of a Micro focus X-ray computer tomograph CT-ALPHA (ProCon X-Ray, Germany) equipped with a 160 kV X-ray source and a detector C7942SK-05 (Hamamatsu, Japan) with 2024 x 2024 active pixels. The unit was running with 155 kV and the received voxels had a size of 65 μm. Additionally the deformation of the open cell foam structures with the alumina based active coating as a function of the compressive strain was observed in situ with the aid of Micro focus X-ray computer tomography (CT).

Finally, the micro structural phase evaluation was carried out by means of scanning electron microscopy (SEM) with the implementation of electron backscatter diffraction analyses (EBSD) of as coked fracture surfaces in combination with energy dispersive X-ray spectroscopy (EDS). A metallographic preparation process would destroy the structure of in situ formed nanoscaled shapes. Therefore, the EBSD phase analysis was carried out on fracture surfaces without further preparation. Due to the roughness of a fracture surface, EBSD is only possible in areas where the proper geometrical conditions for EBSD are given. The second precondition is a complete set of all possible phases and their lattice parameters. Table III shows the phases considered in the present study according also to Roungos et al.[14] The investigation was carried out in the scanning electron microscopes Philips ESEM and Philips XL30 equipped with an EBSD system TSL from Edax/Ametek.

Table III. Phases considered in the EBSD phase analysis.

Name	Crystallographic system	a [Å]	b [Å]	c [Å]	γ [°]	Source
Al₃CON	trigonal	5.454	5.454	14.94	120	ICDD Nr. 0481584
α – Al₂O₃	trigonal	4.76	4.76	12.99	120	ICDD Nr. 0893072
C (graphite)	hexagonal	2.47	2.47	6.8	120	ICDD Nr. 0010640

RESULTS AND DISCUSSION

The slurries for spray coating of Table I show a shear thinning behavior whereby the slurry with the Carbores® P, the graphite and the carbon black achieves lower viscosities of approximately 65 mPas at 600 1/s shear rates in comparison to the pure alumina slurry with the additional PPG wetting additive in the range of 100 mPas. According to the strut thickness distribution due to CT measurements, three areas can be identified of the "Carbores Filter" composition as listed in Table I. Two approximately equal areas of 8mm thickness on the top and on the bottom of the filter with strut thicknesses of 400μm (±150μm) and a middle area of 6mm with strut thicknesses of 300μm (±100μm).

All filters have survived the "small impingement" test, whereby the filters without an alumina based active coating present cracks at the sharp edges of the burned out polyurethane foam, Figure 2. Figure 3 presents a filter with an alumina based active coating after an impingement test, whereby micro cracks are generated in the body of the struts and are stopped also in the body. No critical micro cracks can be identified in the area of the sharp edges. Based on Figure 3 and 4 it is clearly that no chemical bonding between the alumina coating and the alumina carbon bonded substrate exists. In spite this fact due to the higher shrinkage of the alumina coating during coking at 1400°C (see Figure 5), the alumina coating remains on the surface of the "Carbores Filter" after coking as well as after impingement test and contributes with compression stresses especially in the area of the sharp edges. This could be an explanation of the micro crack generation in the body and not at the sharp edges of the burned out polyurethane foam. Additionally in Figure 3 the deviation of the thickness of the coating can be registered as a function of the spraying position of the filter. In the direction of spraying approximately

80μm thickness of the alumina coating can be identified; in the opposite in the area behind only 20μm can be registered. In spite this deviation of the thickness, the alumina coating remains on the filter during the impingement test and is removed after the steel melt is frozen due to a mechanical mainly interlocking with the frozen metal front.

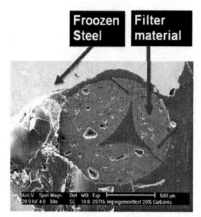

Figure 2. SEM micrograph of a carbon bonded filter with 20 wt% Carbores® P (Carbores Filter) after "small impingement" test, 5kg steel melt 42CrMo4 at 1670°C.

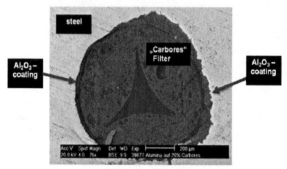

Figure 3. SEM micrograph of a carbon bonded filter with 20 wt% Carbores® P (Carbores Filter) with an alumina based active coating after "small impingement" test, 5kg steel melt 42CrMo4 at 1670°C.

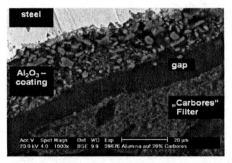

Figure 4. SEM micrograph of a carbon bonded filter with 20 wt% Carbores® P (Carbores Filter) with an alumina based active coating after "small impingement" test, 5kg steel melt 42CrMo4 at 1670°C, magnification at the interface frozen steel melt/alumina coating/carbon bonded filter.

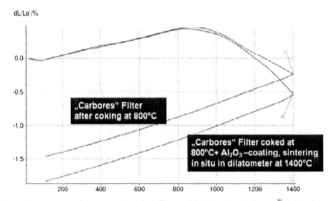

Figure 5. Dilatometer curves of a carbon bonded filter with 20 wt% Carbores® P (Carbores Filter) with and without an alumina based active coating.

An example of the survived impingement test of the carbon bonded filters is given in Figure 6. Related to the mercury porosimetry measurements the open porosity of the struts of the "Carbores Filter" as listed in Table I reach values in the area of 30vol.%. At this point we should underline, that Carbores® P on the one hand is melting at approximately 238°C and is infiltrating the pores of the matrix (this contributes to improved mechanical properties), on the other hand 15vol.% of the Carbores® P amount are organics that produce open pores during coking. Figure 7 shows a mostly dense surface layer of the alumina based active coating.

Figure 6. Photograph of a carbon bonded filter with 20 wt% Carbores® P (Carbores Filter) with an alumina based active coating after "small impingement" test, 5kg steel melt 42CrMo4 at 1670°C.

Table IV. Properties of "Carbores Filters" with and without an alumina based active coating (20 filters have been examined for each composition).

	10 ppi	Weight (without alumina coating) (gr.)	Weight (with alumina coating) (gr.)	Strut thickness Carbores filter/Alumina coating (μm)	CCS (MPa)
Coked at 800 °C and 1400 °C	Carbores filter	19.5± 0.2		300/0	0.33 ± 0.04
	Carbores filter + thin alumina coating	19.5± 0.2	26.6± 0.1	300/80	1.11± 0.16

Figure 7. SEM micrograph of a mostly dense surface layer of the alumina based active coating.

In Table IV the CCS strengths related to the weight and the strut thicknesses are listed. It is obvious that the alumina based active coating contributes to a higher strength. Observing the CT-images due to the in situ compression in the CT we can assume that the alumina coating is not breaking out of the carbon bonded substrate, Figure 8 and 9. This can also be observed in the fragments of the filter after the CCS-test, Figure 10.

Table V shows the strengths of the three different versions of the nano-engineered filter compositions. The combination of the two types of nano-scaled additions led to high mechanical strengths accompanied by low deviations. With the aid of the EBSD the Al_3CON phase has been identified, Figure 11 and 12. With respect to the in situ formation of the new Al_3CON phase, the higher mechanical strengths are attributed to the chemical interconnecting phase of Al_3CON, which is assumed that contributes to a bonding between the oxide alumina filler and the carbon. Similar results have been reported by Roungos and Aneziris.[14]

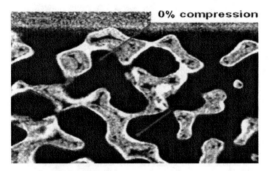

Figure 8. CT-image of a carbon bonded filter with 20 wt% Carbores® P (Carbores Filter) with an alumina based active coating, 0% compression.

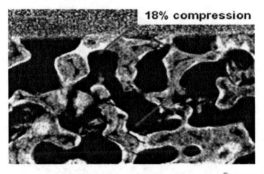

Figure 9. CT-image of a carbon bonded filter with 20 wt% Carbores® P (Carbores Filter) with an alumina based active coating, 18% compression.

Figure 10. Fragments of a carbon bonded filter with 20 wt% Carbores® P (Carbores Filter) with an alumina based active coating after CCS test.

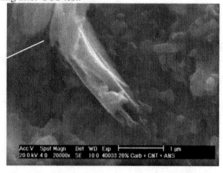

Figure 11. Al₃CON phase in a "Carbores Filter" with nanoscaled additives based on CNTs and alumina NS.

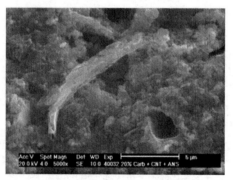

Figure 12. Al₃CON phase in a "Carbores Filter" with nanoscaled additives based on CNTs and alumina NS.

Table V. Strengths of nano-engineered filters.

	10 ppi	Weight (gr.)	CCS (MPa)	Strength increase (%)
	Carbores Filter	19.5± 0.2	0.33 ± 0.04	0
Coked at 1000 °C	Carbores Filter, only spray slurry with CNT+ Alumina NS	19.5± 0.2	0.46 ± 0.11	25
	Carbores Filter, imregnation + spray slurry with CNT+ Alumina NS	19.5± 0.2	0.58± 0.06	43
	Carbores Filter, imregnation + spray slurry with CNT	19.5± 0.2	0.8 ± 0.40	58

CONCLUSIONS

New filter approaches related to new carbon bonded compositions with and without nanoscaled additives have been demonstrated. The reinforcing of the carbon bonded compositions due to the nanoscaled additives is a first step for filter production with higher filtration capacities in means of bigger dimensions. The alumina active based coating improves the mechanical properties and will be checked in the future according to its filtration efficiency performance at the CRC in Freiberg as well as in industrial oriented steel castings in cooperation with foundries.

REFERENCES

[1]C. Klinger, D. Bettge, R. Häcker, Schadensanalyse Radsatzwelle ICE 3, Vorab-Präsentation EBA, DB, 24.09.2008, Bundesanstalt für Materialforschung und –prüfung (2008)
[2]L. Zhang, B. G. Thomas, Inclusions in Continous Casting of Steel, *Proceedings* of XXIV National Steelmaking Symposium, Morelia, Mich, Mexico, 26-28, 138-183 (2003)
[3]K. Wasai, K. Mukai, A. Miyanaga, Observation of Inclusion in Aluminum Deoxidized Iron, *ISIJ International*, Vol. 42, No. 5, 459-466(2002)
[4]S. Ovtchinnikov, Kontrollierte Erstarrung und Einschlussbildung bei der Desoxidation von hochreinen Stahlschmelzen, *Dissertation Thesis* TU Bergakademie Freiberg, (2002)

[5]C.G. Aneziris, A. Ansorge, H. Jaunich, New approaches of carbon bonded foam filters for filtration of large castings. *Cer. Forum Intern.*, 85 No.10, E100-E103(2008)

[6]D. Janke, K. Raiber, Grundlegende Untersuchungen zur Optimierung der Filtration von Stahlschmelzen, *Technische Forschung Stahl*, Luxembourg Europäische Kommission, - ISBN 92-827-6458-3, (1996)

[7]P. Hammerschmid, D Janke, Kenntnisstand zur Abscheidung von Einschlüssen beim Filt-rieren von Stahlschmelzen, *Stahl und Eisen,* 108 Nr.5, 211-219 (1988)

[8]O. Davila-Maldonado, A. Adams, L. Oliveira, B. Alquist, R.D. Morales, Simulation of fluid and inclusions dynamics during filtration operations of ductile iron melts using foam filters, *Metallurgical and Material Transactions B* 39, 818-839 (2008)

[9]K. Schwartzwalder, Method of making porous ceramic articles, *Patent US 3090094* (1963)

[10]X. Zhu, D. Jiang, S. Tan, The control of slurry rheology in the processing of reticulated porous ceramics, *Materials Research Bulletin*, 37, 541-553 (2002)

[11]J. Saggio-Woyansky, C.E. Scott, W.P. Minnear, Processing of Porous Ceramics, *American Ceramic Society Bulletin*, 71, 11, 1674-1682 (1992)

[12]M. Hasterok, C. Wenzel, C.G.Aneziris, U. Ballaschk, H. Berek, Processing of ceramic preforms for TRIP-matrix-composites, *Steel Research International*,82(9),1032-1039 (2011)

[13]V.N. Antsiferov, S.E. Porozova, Foam ceramic filters for molten metals: reality and prospects. *Powder Metallurgy and Metal Ceramic*, 42, 9-10 (2003)

[14]V. Roungos, C.G. Aneziris, H. Berek, Novel Al$_2$O$_3$-C refractories with less residual carbon due to nanoscaled additives for continuous steel casting applications, *Advanced Engineering Materials*, DOI: 10.1002/adem.201100222, accepted, article first published online 24 NOV. 2011

ACKNOWLEDGEMENTS
This work was financially supported by the German Research Foundation (DFG) in frame of the Collaborative Research Center 920.

FAILURE AND STIFFNESS ANALYSIS OF CERAMIC FROM A 25-mm DIAMETER DIESEL
PARTICULATE FILTER

Ethan E. Fox, Andrew A. Wereszczak, Michael J. Lance, and Mattison K. Ferber
Ceramic Science and Technology
Oak Ridge National Laboratory
Oak Ridge, TN 37831-6068

ABSTRACT

Three established mechanical test specimen geometries and test methods used to evaluate mechanical properties of brittle materials are adapted to the diesel particulate filter (DPF) architecture to evaluate failure initiation stress and apparent elastic modulus of the ceramics comprising DPFs. The three custom-designed test coupons are harvested out of 25-mm diameter DPFs to promote a particular combination of orientation of crack initiation and crack plane. The testing of the DPF biaxial flexure disk produces a radial tensile stress and a crack plane parallel with the DPF's longitudinal axis. The testing of the DPF sectored flexural specimen produces axial tension at the DPF's OD (outer diameter) and a crack plane perpendicular to the DPF's longitudinal axis. The testing of the DPF o-ring specimen produces hoop tension at the DPF's original OD and at the inner diameter of the test coupon, and a crack plane parallel to the DPF's longitudinal axis. The testing of these mechanical test coupons also enables the determination of a secant elastic modulus of the DPF ceramic material. Results consistently show that the secant elastic modulus of the DPF ceramics at tensile failure strain is approximately 1-2 GPa; this is approximately one order of magnitude less than the apparent elastic modulus estimated using sonic- or resonance-based test methods. However, the estimated tensile failure stresses are equivalent to those reported in other studies. The lower elastic modulus estimated in this work means that predicted tensile stresses in DPFs will be lower and that lifetime should be higher compared to when misleadingly high, sonic-based modulus values are used in those analyses.

1. INTRODUCTION

Diesel particulate filter (DPF) technology enables the fuel efficient diesel engine to meet emission regulations for particulate matter. An example of a DPF is shown in Fig. 1. The DPF collects particulate from the exhaust stream during operation. It then is periodically regenerated by increasing its temperature to nominally 600°C. This causes the carbon-containing trapped particulate to oxidize and form gaseous CO_2 enabling continued filtration at a lower backpressure. But the regeneration causes thermal gradients in the brittle ceramic which can be problematic.

Two lifetime-limiting consequences of overly severe thermal gradients are crack initiation and propagation. The end crack and ring crack are examples of crack types that form in DPFs and are schematically shown by the black lines in Fig. 1. The formation and existence of either can compromise the intended filtering function of the DPF and ultimately result in their need for replacement. Therefore, the end-user has interest in managing the operating conditions so the DPF successfully fulfills its purpose without initiating those cracks.

Two criteria must be satisfied in tandem to promote and achieve maximum service lifetimes in DPFs under severe operating conditions. The ceramic comprising the DPF must have sufficiently high tensile strength and the tensile stresses (such as those caused by operating thermal conditions) must be sufficiently low. Obviously, if the anticipated operating conditions produce high tensile stresses, then

139

a stronger DPF should be considered for use. Or if the only available candidate DPF ceramics have a limited tensile strength, then the operating conditions need to be restrained so that too-high tensile stresses do not arise. In either circumstance, the understandings of both the DPF ceramic strength and the DPF service conditions are needed to successfully employ a DPF.

The developed service stress state in a DPF can be complex owing to its anisotropic architecture. The DPFs examined in this study have a 4-fold rotational symmetry (C4). The thermal gradients (and consequential mechanical stresses) for operation will of course be affected by this anisotropic structure, and out of convenience may be interpreted with respect to cylindrical coordinates; namely, axial (i.e., gas flow direction), radial, and hoop orientations.

Figure 1. Schematics of (top) end crack and (middle) ring crack superimposed on a DPF. The plane of the end crack is parallel to the DPF's longitudinal axis and intersects the DPF's end face. The ring crack is perpendicular to the DPF's longitudinal axis and intersects the outside diameter usually near the middle of the DPF's length.

Ceramics almost always crack because of the existence of a too-high (First Principal) tensile stress, and the formation of the two crack types schematically shown in Fig. 1 may be described in context to that and the DPF's architecture. The plane of the end crack is axial or parallel to the gas flow direction and intersects the DPF's end. The end crack forms as a consequence of sufficiently high radial tensile stresses existing at the DPF end. The plane of the ring crack is perpendicular to the gas flow direction and intersects the outside diameter of the DPF. Its location is usually near the middle of the axial length and forms as a consequence of sufficiently high axial tensile stresses existing at the exterior of the DPF. In general, when ceramics crack in the presence of a thermal gradient (particularly when there is concurrent confinement), crack initiation usually occurs near the cold side of that gradient and often when there is a sufficiently rapid heating or cooling transient (i.e., thermal shock).

To successfully predict thermomechanical stresses (and predict crack or failure initiation), the thermoelastic properties, elastic properties, and tensile failure strength of the DPF ceramic must be known or measured.

While the porous microstructure of DPF ceramics is a requirement for exhaust gas filtering it also produces a complex response when subjected to (thermo) mechanical loading. The interior struts are open and continuous (filtering-capable) while the exterior skin is less porous and closed. Such porous structures often have an asymmetric response to tensile and compressive loadings. Tensile loading of the compliant porous structure can readily cause initiation, coalescence, and accumulation of local microcracking resulting in a gradual compliance increase with additional tensile straining. While compressive loading will also cause microcracking, additional compressive straining beyond its initiation can lead to little change in compliance with additional microcracking. The tensile failure stress, like for almost all ceramics, has a much lower magnitude than compressive failure stress, so rationale exists to focus attention on the tensile failure stress for appropriate and conservative DPF reliability analysis.

There were two primary goals to this study. The first was to employ test specimens whose mechanical loading would produce tensile-stress-induced failure modes consistent with those of the crack types shown in Fig. 1 and that would enable quantification of tensile failure stress associated with their formation. Flexure testing works well for achieving that and tends to provide greater ease of valid experimentation than uniaxial tension testing does, and also is readily adaptable to high temperature testing too. Test specimen geometries used in the mechanical evaluation of other types of ceramics were adapted to the mechanical testing of the DPF architecture and ceramic to better facilitate the understanding of the end and ring crack in Fig. 1. The second goal was to also use analyses that enable the estimation of the DPF ceramic's elastic modulus - a property needed for any eventual modeling of the DPF's operational stress state. The geometrical characteristics of this filter are as follows: nominal diameter of 25mm, length of 75mm, 256 cells per square inch and a manufacturer reported porosity of 59 percent.

2. THREE ALTERNATIVE SPECIMENS

The descriptions of the equibiaxial flexure specimen, sectored flexure specimen, and the o-ring flexure specimen are presented along with their mechanical testing and respective analyses used to determine elastic modulus and tensile failure stress.

2.A. APPROACH

The method to estimate elastic modulus and tensile failure stress with all three alternative test specimens is represented by the flowchart in Fig. 2. The method combines analysis (either analytic or finite element) with mechanical testing. The mechanical testing of all the test coupons produces a record of compressive force (all three test methods are flexure tests and compressive loading coupled with appropriate fixturing activates their flexure) and an associated linear specimen deflection up through a maximum compressive force where crack initiation begins. All three specimen geometries responded linear elastically up to that maximum compressive force. The coupled pair of maximum-compressive-force and deflection data is recorded for every specimen after the important correction (subtraction) for machine compliance. The geometry of that actual specimen is then modeled using FEA (equibiaxial specimen and o-ring specimen) or classical beam bending (sectored flexure specimen) and displacement as a function of compressive force and elastic modulus is predicted. Elastic modulus is then iteratively varied until a produced force-displacement couple finally matches the force-displacement couple measured from the flexure testing. That elastic modulus (secant elastic modulus actually) is then used again in the FEA to estimate the maximum tensile failure stress in that specimen when failure initiation began.

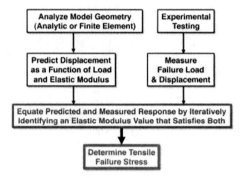

Figure 2.   Analysis of all three alternative test coupons were combined with their experimental testing to (1) estimate secant elastic modulus of the DPF cordierite and (2) failure stress.

2.B. EQUIBIAXIAL FLEXURE (RADIAL TENSION)

The testing of the equibiaxial flexure specimen subjects the DPF's end structure (or the interior strut structure if the disk is cut from the interior) to a biaxial radial tensile stress, and its measured response is relatable to the interpretation of end-crack formation illustrated in Fig. 1.

The equibiaxial flexure specimen and test are a standard method (ASTM C1499) used for measuring failure stress in dense ceramic materials [1] and it was adapted here to measure radial tensile failure stresses in DPFs. An advantage of its use with disks harvested out of a DPF is its produced biaxial flexure stress state will result in failure of the weakest direction in the biaxially stressed plane as compressive force is increased. Examples of the equibiaxial flexure specimens, tested coupons, and load and support rings are shown in Fig 3. A coupon set up in the test fixture is shown in Fig. 4. The geometry, applied compressive force, and resulting disk deflection for the equibiaxial flexure specimen (a.k.a., ring-on-ring or RoR specimen) were modeled in 3D using ANSYS (see Fig. 5), and their predicted load-displacement responses were matched with RoR flexure testing results using the method shown in Fig. 2 to estimate radial tensile failure stress and the material's secant elastic modulus.

Disk size and RoR span diameters were chosen so to produce classical beam bending (i.e., deflection to failure less than ~ 1/4 the specimen thickness) while avoiding the application of high contact stresses from the rings (i.e., avoid contact-induced crushing). All equibiaxial flexure disks responded linear elastically. Failure forces for all specimens were relatively low-valued and there was never evidence seen that the rings had caused contact-induced crushing. Test fixture rings were made of acetyl and an example of one paired size is shown above the coupons in Fig. 3. The rings could alternatively be made out of high-temperature-capable material if high-temperature testing of the DPF was to be sought.

Figure 3. Load and support rings along with a fractured end disk, and untested center, and end disks.

Figure 4. An equibiaxial disk setup in the test fixture.

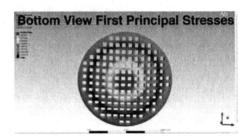

Figure 5. Bottom view (tensile surface) of equibiaxial disk subjected to flexure. FEA was combined with experimental measurement to estimate secant elastic modulus and failure stress.

2.C. SECTORED FLEXURE (AXIAL TENSION)

The testing of the sectored flexure specimen subjects the DPF's outer diameter (or exterior skin) to an axial tensile stress and its measured response is applicable to the prediction and the interpretation of ring crack formation illustrated in Fig. 1.

The sectored flexure specimen was developed to measure the axial tensile failure stress in dense ceramic tubes and to harvest and test numerous specimens from each tube to improve statistical confidence [2]. Portions of its machining are illustrated in Fig. 6. Sectors are first axially sectioned out of the DPF. An isosceles triangle is then cut from the sector to produce a "sectored flexure specimen" with a flat on one side and the original outer diameter of the DPF retained on the opposite. The specimen is oriented in the flexure fixture so its flat side is the compressive side and the opposite curved portion is tensile loaded. The nominal dimensions of the sectored flexure specimens are as follows: 75mm long, nominally 4 mm thick and 12 mm wide.

A top view along with an end view of the sectored flexure specimens is shown in Fig. 7 and a coupon setup in a test fixture is shown in Fig. 8. Load and support rings are 6.75-mm and 19-mm respectively. For the 4-pt bend fixture shown in Fig. 8, hour-glass shaped rollers were used for the bottom and cylindrical rollers for the top. The radius of curvature for the hour-glass shaped rollers matched that of the DPF's original outside diameter so to promote line loading along its curved surface so to minimize contact loading stresses between the roller and specimen. Cylindrical rollers could be used on the fixture's top because it was a flat surface. Load and support spans for the 4-pt bend are 30-mm and 60-mm respectively. Specimens failed at sufficiently low forces to sustain classical beam bending (when coupled with the chosen upper and lower fixture span sizes and specimen's cross-section or moment of inertia, MOI). Additionally, all sectored flexure specimens responded linear elastically up to their failure force. Loading displacement rate for biaxial flexure and sectored flexure testing was 0.5 mm/min.

The geometry, applied compressive force, and resulting specimen deflection for the sectored flexure specimen were modeled using Oak Ridge National Laboratory- (ORNL)-developed software that calculates the MOI and centroids for bend specimens with complicated cross-sections comprising dissimilar materials. A digital photograph of the specimen cross-section is used as input. The user specifies dimensional-scale and the elastic properties of the different materials in the photo represented by different gray-scales. The software then calculates the MOI and distance from outer fiber to the neutral axis. Their values are then used in the classical beam bending equation along the fixture spans and failure force to calculate the axial tensile failure stress. And to estimate elastic modulus of the structure, the predicted load-displacement bending responses were matched with sectored flexure testing results using the method shown in Fig. 2.

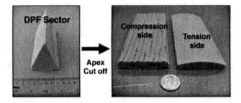

Figure 6. A sector is cut from a DPF and then an isosceles triangle is cut to produce a sectored flexure specimen. The DPF's original outer diameter is also the tensile surface during flexure testing.

Figure 7. Bottom and end views of sectored flexure specimens.

Figure 8. Example of a sectored flexure specimen positioned in the 4-pt bend fixture. Upper and lower spans are 30 mm and 60 mm respectively.

2.D. O-RING FLEXURE (HOOP TENSION)

The o-ring specimen was developed to estimate hoop tensile failure stresses on the ID and OD of dense ceramic tubes [3-4]. Likewise, the testing of the adapted o-ring specimen subjects the DPF's outer diameter (or skin) to hoop tensile stress. The nominal dimensions for the o-ring are 5 mm thickness, 16 mm inner diameter and a 25 mm outer diameter. The OD is as received from the manufacturer.

An image of an o-ring specimen in the test fixture alongside a schematic diagram of a loaded specimen is shown in Figure 9. The specimen is diametrally compressed, and all DPF o-ring specimens responded linear elastically up to some maximum force where a crack initiates. The produced (First Principal) tensile stress field for the diametrally compressed o-ring specimen is shown on the left in Fig. 9. Regions of maximum hoop tension are produced at the 3 and 9 o'clock positions at the OD and at the 6 and 12 o'clock positions at the ID. Postmortem of the failed o-ring specimen identifies where fracture had commenced.

The geometry, applied compressive force, and resulting o-ring deflection were modeled in 2D using μ-FEA and ANSYS, and their predicted load-displacement responses were matched with measured o-ring flexure testing results using the method shown in Fig. 2. This allowed estimation of the secant elastic modulus and hoop tensile failure stress at the exterior skin or in the strut structure (depending on where failure initiation had occurred in the o-ring specimen).

Though the produced crack plane from the o-ring specimen is not linkable to either of the two crack types shown in Fig. 1, its testing enables the measurement of hoop tensile stress and this in turn allows the potential concurrent prediction and consideration of what operational stress states in a DPF would activate such a crack type. The plane of such a produced crack would be parallel to the DPF's flow direction or axis of symmetry and intersect the outer surface to produce a crack visible from the outside or intersect the interior of the skin (which would not be seen from the outside).

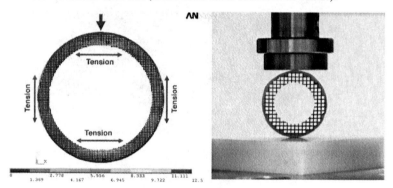

Figure 9. Examples of FEA modeling (left) and an o-ring specimen in a test fixture.

3. CORRELATING STRAIN AND STRESS

If a material is symmetrically linear elastic in both the tension and compression domains, then the applied stress and strain are simply related by the material's elastic modulus through the classical Hooke's Law. However, as discussed in the Introduction, the porous ceramics comprising DPFs are likely not symmetric in the tension and compression domains. Additionally, there is ample evidence to show that these materials are not linear either. In other words, *the stress-strain response in these porous materials are likely asymmetrically non-linear.* Furthermore, they are known to exhibit hysteresis in cyclic loading which further complicates the understanding of their mechanical response; but in this study, our focus is only on the state of the material up to the (tensile) strain or stress where damage initiates. All these contributing complications have led to misplaced focus and improper interpretation of how the ceramics in DPFs mechanically respond.

Ideally, the stress-strain response would be available over a large domain of compressive and tensile strains. It is not. Given that the magnitudes of tensile failure stress will be lower than those of compressive failure stresses, an attractive consolation would be if there was an understanding of the stress-strain response in the tensile strain domain. There is nothing for that either.

Why cannot a DPF be conventionally strain-gaged and uniaxially tensile stressed to measure a tensile stress-strain response? The porous and weak structure of the DPF cannot accommodate strain-gage mounting nor permit gripping of a tensile fixture. Non-contact extensometry would not necessarily solve this problem because the tensile gripping difficulty would still be an impassable obstacle. Unfortunately, there indeed is no elegant or conventional method available to measure a continuous, tensile stress-strain response in a DPF ceramic.

To start to estimate an elastic modulus of porous DPF structures, many have resorted to the use of sonic-based or resonance-based measurements [5]. These techniques rely on infinitesimally small excitations of displacements or strains for their measurement. They are recognized to be reliable test methods for validly measuring this material property when the material is symmetrically linear elastic over a large domain of compressive and tensile strain. However, there is a problem with that whose description can be facilitated by what is illustrated in Fig. 10. If a material is asymmetrically non-linear in the compressive and tensile strain domains, and that asymmetry exhibits itself at small absolute values of strains, then a sonic- or resonance-based measurement of elastic modulus is only meaningful at those very small strains where the local stress-strain response is symmetrically linear elastic. The elastic modulus measured at those very small strains is not going to be representative of, and of course not equal to, the (tangent) elastic modulus at tensile (or compressive) strains of DPF operational service. They will be higher valued, and the results from the present study support that. Such a difference in dynamic and quasi-static elastic modulus has precedence - thermal barrier coatings exhibit an analogous difference [6-7] and have a mechanical asymmetry.

In *every* case where we have estimated elastic modulus with the specimens described in this study using the method shown in Fig. 2, we have determined a (secant) elastic modulus, E_{SEC}, that is 4-12 times *lower* than reported E's estimated by sonic- or resonance-based test methods. There have been no outliers whose values approach the values measured by sonic- or resonance-based methods. This comparison is summarized in Table 1 and although the values listed are a function of porosity they still provide a course comparison. Our employed method is actually determining the E_{SEC}; which actually is still going to be higher-valued than the tangent modulus, E_{TAN}, for any given tensile strain. We are producing tensile failure at strains in the range of 0.1-0.3% in all three of these specimens (i.e., failure at strains that are many orders of magnitude higher than those that are associated with sonic- or resonance-based test methods). Despite that, the E_{SEC} is still a more representative value to report than that measured by sonic- or resonance-based methods.

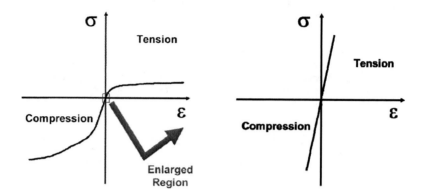

Figure 10. Sonic- or dynamic-based elastic modulus measurement methods evaluate stiffness at extremely small displacements and will therefore overestimate elastic modulus for non-linear materials.

Uniaxial compression testing is another option that has been pursued for estimating elastic modulus. Bruno et al. [8] show a trend of elastic modulus decrease as a function of compressive stress as measured by neutron diffraction. Zero strain or stress values for elastic modulus were not reported; however, the highest elastic modulus values are at low compressive stress and are approximately 13 GPa and decrease to approximately 6 GPa at approximately 9 MPa of compressive loading (included for comparison in Table 1). These trends are consistent with the compressive strain domain in the Fig. 10 schematic; however, because of the stress-strain asymmetry in the compressive and tensile strain domains, Bruno et al. reported elastic modulus values will be overestimates of elastic modulus in the tensile strain domain and the same absolute values of strain.

What will happen if sonic- or resonant- or dynamic-based measurements of elastic modulus or if compression-loading-based measurement of elastic modulus are used in the design or reliability analysis of an operating DPF? Their large-values will predict erroneously high tensile stresses and erroneously short lifetimes. The quasi-static or mechanical-based measurements featured in this study are more appropriate for use (in the continued absence of a continuous stress-strain curve in both the compressive and especially the tensile strain domains).

The estimation of the E_{SEC} using the algorithm shown in Fig. 2 was the first of two informative outcomes of this study; the estimation of tensile failure stress (for different orientations) was the other. Using our E_{SEC} values from Table 1 back in our various analyses for the three specimens (i.e., FEA or MOI analysis), we are able to estimate the tensile stress failure for all specimens. There is a suggestion that the axial tensile failure stresses for the exterior skin are higher valued, but that work is continuing and we look to confirm that with additional testing. See Table 2 for a comparison of the tensile failure stresses of each of the three coupon geometries.

While the focus of this paper has been on 25-mm-diameter cordierite DPFs, analogous results have been generated by the authors with both 143-mm- and 230-mm-diameter DPFs. Namely, the elastic modulus found using the algorithm pictured in Fig. 2 was on the order of 1-3 GPa for those two larger diameters [9] while the elastic modulus reported by the manufacturer (using sonic or resonance methods) is between 4 and 13 GPa. These results are very consistent for all three DPF diameters which demonstrate their independence of DPF size.

Table 1. Comparison of elastic moduli.

Test Method	Elastic Modulus (GPa)*	Source
Dynamic: Resonance-Based	4 - 7	[10]
Dynamic: RUS	12.3 ± 0.3	[5]
Quasi-static / Mechanical Uniaxial Compression	> 13 GPa at zero MPa; ~ 6 GPa at 9 MPa	[8]
Quasi-static / Mechanical Equibiaxial Flexure	0.5 - 1.5	Present study
Quasi-static / Mechanical Sectored Flexure	1 – 3 (Interior strut) 4 – 24 (Exterior skin)	
Quasi-static / Mechanical O-Ring Flexure	1.1 - 2.1	

   * A function of porosity - comparison intended to illustrate coarse
     comparison.

Table 2. Comparison of quasi-static tensile failure stresses.

Test Method	Tensile Failure Stress (MPa)*	Source
Equibiaxial Flexure	2 (est) (interior strut) > 4 (est) (50% end fill) radial	Present study
Sectored Flexure	5 - 13 (est) (exterior skin) axial	
O-ring Flexure	2 - 4 (est) (exterior skin) hoop	

   * A function of porosity - comparison intended to illustrate coarse
     comparison.

For our future work, we will continue to test these specimens to generate strength statistics and also attempt to develop additional modifications of the three specimens so that we can produce failure at different tensile strains, and start to populate or infer a tensile-stress-strain diagram for porous ceramics for DPFs. Also, we are working to harvest and isolate specimens from the interior struts, the exterior skin, and from the 50% end fill so that we may mechanically measure the elastic modulus of all three, and take them all into account in future operational stress modeling of DPFs. Preliminary bend tests on stand-alone struts cut from DPFs, where both applied force and responding displacement were measured, show that the elastic modulus of the DPF ceramic is approximately 1 GPa, which is equivalent to the secant modulus measured with the herein described three test coupons.

4. CONCLUSIONS

An E_{SEC} in the range of 1 - 2 GPa consistently results in analytical (FEA or numerical) and experimental correlation for deformations of DPF cordierite ceramic equibiaxial flexure, sectored flexure, and o-ring flexure specimens and for three different DPF diameters. That range of E_{SEC} is about 4-12 times lower than reported elastic modulus for cordierite ceramics sonically and dynamically measured. If this lower E_{SEC} is correct, then actual stresses in a DPF during service will be much lower than that predicted using sonically or dynamically measured elastic modulus. Owing to the probable non-linearity and asymmetry of stress and strain with these DPF cordierite microstructures, a strain-to-failure criterion may be more relevant for use than a failure stress criterion.

5. REFERENCES

1. ASTM C1499, "Standard Test Method for Monotonic Equibiaxial Flexural Strength of Advanced Ceramics at Ambient Temperature," Vol. 15.01, ASTM International, West Conshohocken, PA, 2009.
2. Wereszczak, A. A., Duffy, S. F., Baker, E. H., Jr., Swab, J. J., and Champoux, G. J., *Journal of Testing and Evaluation*, 2008, 36:17-23.
3. Jadaan, O. M., Shelleman, D. L., Conway, J. C. Jr., Mecholsky, J. J. Jr., and Tressler, R. E., "Prediction of the Strength of Ceramic Tubular Components: Part I - Analysis," *Journal of Testing and Evaluation*, 1991, 19:181-91.
4. Shelleman, D. L., Jadaan, O. M., Conway, J. C., Jr., and Mecholsky, J. J. Jr., "Prediction of the Strength of Ceramic Tubular Components: Part II - Experimental Verification," *Journal of Testing and Evaluation*, 1991, 19:192-200.
5. Shyam, A., Lara-Curzio, E., Watkins, T. E., and Parten, R. J., *Journal of the American Ceramic Society*, 2008, 91:1995-2001.
6. Eldridge, J. J., Morscher, G. N., and Choi, S. R., "Quasistatic vs. Dynamic Modulus Measurements of Plasma-Sprayed Thermal Barrier Coatings," *Ceramic Engineering and Science Proceedings*, 2002, 23:371-78.
7. Choi, S. R., Zhu, D., and Miller, R.A., "Effect of Sintering on Mechanical Properties of Plasma-Sprayed Zirconia-Based Thermal Barrier Coatings," *Journal of the American Ceramic Society*, 2005, 88:2859-67.
8. Bruno, G., Efremov, A. M., Levandovskiy, A. N., Pozdnyokova, I., Hughes, D. J., and Clausen, B., "Thermal and Mechanical Response of Industrial Porous Ceramics," *Materials Science Forum*, 2010, 652:191-96.
9. Wereszczak, A. A., E. E. Fox, M. J. Lance, and M. K. Ferber, "Failure Stress and Apparent Elastic Modulus of Diesel Particulate Filter Ceramics," *to be given at SAE International World Congress*, Detroit, MI, April 2012.

10. Kuki, T., Miyairi, Y., Kasai, Y., Miyazaki, M. and Miwa, S., "Study on Reliability of Wall-Flow Type Diesel Particulate Filter," SAE Paper Number 2004-01-0959, SAE International, 2004.

6. CONTACT INFORMATION

Andrew A. Wereszczak, wereszczakaa@ornl.gov, 865.576.1169 (voice).

7. ACKNOWLEDGMENTS

Research sponsored by the U.S. Department of Energy, Assistant Secretary for Energy Efficiency and Renewable Energy, Office of Vehicle Technologies, as part of the Propulsion Materials Program, under contract DE-AC05-00OR22725 with UT-Battelle, LLC. The authors thank G. Gonze and J. Tan of General Motors for their input and J. Parks and R. Wiles of ORNL for their technical reviews of the manuscript and suggestions.

DEVELOPMENT OF POROUS SiC WITH TAILORABLE PROPERTIES

Prashant Karandikar, Glen Evans, Eric Klier*, and Michael Aghajanian
M Cubed Technologies, Inc.
1 Tralee Industrial Park
Newark, DE 19711
*Currently at ARL, Aberdeen, MD 21005

ABSTRACT

Silicon carbide offers high stiffness, light weight, high hardness, high thermal conductivity, low thermal expansion, high wear resistance, and high corrosion resistance at room as well as elevated temperatures. Hence it is used in applications such as armor, industrial wear components, chemical processing, mining, and semiconductor processing. With the above attributes, porous SiC can be used in applications such as filters (diesel particulate, coal, oil, gas, separation membranes), catalyst supports, substrates, sensors (relative humidity, NH_3) etc. In this study, porous SiC was made by two processes, reaction bonding and pressureless sintering. Process variables such as raw materials, infiltration approach, temperature, time, etc. were systematically varied. Microstructure, porosity content, pore size distribution, average pore size, and permeability of the resultant materials were characterized. The ability to make parts with a combination of micro-porosity and macro porosity (channels) was demonstrated.

INTRODUCTION

Silicon carbide offers high stiffness, light weight, high hardness, high thermal conductivity, low thermal expansion, high wear resistance, and high corrosion resistance at room as well as elevated temperatures. Hence it is used in applications such as armor, industrial wear components, chemical processing, mining, and semiconductor processing.

SiC can also be made in the porous form. Typically, two regimes of porosity can be found, low (up to 50%) and high (up to 90%). The porous ceramics in the second regime are typically referred to as "cellular" ceramics or ceramic "foams". The special attributes described above make porous SiC a prime candidate in applications such as filters (diesel particulate, coal, oil, gas, separation membranes), catalyst supports, substrates, burners, sensors (relative humidity, NH_3), etc. [1-7]. In most applications porous SiC is used in the bulk form. In the sensor applications, it is used as a porous film.

PROCESSING METHODS

Most standard processes for making conventional SiC can be adapted to make porous SiC. These include

- Reaction bonding
- Pressureless sintering
- Pre-ceramic polymer infiltration and pyrolysis
- Vapor deposition/patterning
- Selective etching

The characteristics of porous SiC made are affected by the process selected for their manufacture and the following process parameters:

153

- Starting particle size
- Binders
- Process temperature and time
- Additives

In this work reaction bonding and pressureless sintering were used to make porous SiC coupons. These processes are described below.

Reaction Bonding

In the reaction bonding process (silicon-based matrices) [8-9], good wetting and highly exothermic reaction between liquid silicon and carbon is utilized to achieve pressure-less infiltration of a powder preform. This process is given many names such as reaction-bonding, reaction-sintering, self-bonding, and melt infiltration. A schematic of the reaction bonding process is shown in Figure 1. The steps in the process are as follows (1) Mixing of SiC powder and a binder to make a slurry, (2) Shaping the slurry by various techniques such as casting, injection molding, pressing etc., (3) Drying and carbonizing of binder, (4) Green machining, (5) Infiltration (reaction bonding) with molten Si (or alloy) above 1410°C in an inert/vacuum atmosphere, and (6) Solidification and cooling.

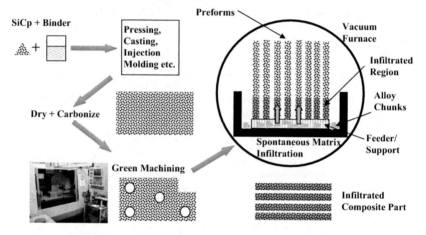

Figure 1. Schematic representation of the reaction bonding process.

Pressureless Sintering

Porous SiC samples were also prepared by the pressureless sintering [10] approach. In this process, a preform of submicron SiC powders and sintering aids is prepared by cold pressing and then heated to high temperatures (>1700°C). The high surface area of these powders provides the driving force for mass transport, grain growth and sintering [10].

EXPERIMENTAL PROCEDURE

The reaction bonding and sintering process parameters were systematically varied to make porous SiC with
- Various total porosity contents
- Different pore sizes
- Different permeabilities

Four different samples were produced by each processing method. These were subjected to physical, microstructural, chemical, and thermal characterization. Small coupons were mounted in epoxy and prepared by the standard metallographic techniques for microstructural examination by optical and scanning electron microscopy (SEM). Fracture surfaces of broken samples were also observed by SEM. For high porosity samples, SEM is the better approach for microstructural examination due to the depth of field available. Chemical composition was analyzed by the GDMS (glow discharge mass spectroscopy) technique.

Densities of the resultant composites were characterized using the Archimedes principle, (ASTM B311). In addition to the dry weight and the suspended weight, "wet" weights of the samples were measured to calculate open porosity. For porous samples, the Archimedes method presents many challenges. Hence, density was also calculated by the geometric method. For this, the length, width, and the thickness of the samples were measured to calculate the volume. Density was calculated by dividing the dry weight by the calculated volume. The geometric density allows calculation of the open as well as closed porosity.

Pore size measurements were made using mercury intrusion porosimetry. In this method, the sample is engulfed in a non-wetting liquid (e.g. mercury). Due to its poor wetting characteristics, the porosity in the sample is not penetrated by mercury at atmospheric pressure. Subsequently, pressure is applied in increments to force the mercury into the pores of the sample. The size (D) of pores penetrated at a given pressure (p) is characterized by the Washburn equation $(D = -4\gamma\cos\theta/p)$ where γ is the surface energy and θ is the contact angle. As the pressure increases, finer and finer pores are intruded by the mercury. The level of mercury in an attached capillary is capacitively monitored to calculate the volume of intruded mercury. Permeabilities were measured by the gas permeability technique (Figure 2). Here, the flow of a gas (e.g. air) through the sample is measured under applied differential pressure. Thermal expansion measurements were made by dilatometry from room temperature to 1350°C.

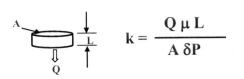

$$k = \frac{Q\,\mu\,L}{A\,\delta P}$$

Gas flow rate - Q: m^3/s
Gas viscosity- μ: Pa s
Length - L: m
Area - A: m^2
Pressure gradient – δP: Pa
Permeability - k: m^2
Darcys: 10^{-12}m^2

Figure 2. Permeability measurements

RESULTS AND DISCUSSION

The properties of samples made by both processing methods with four different porosity levels are summarized in Table I. As can be seen in Table I, total porosity was varied from 6% to 42%. As can be seen from the table, the samples had small amount of closed porosity (total –

open; 0-3%). The samples made by reaction bonding produced coarser porosity. The samples made by sintering produced finer porosity. Example pore size distributions for samples with fine and coarse porosity are shown in Figure 3. The average pore diameters for all samples are also listed in Table I. As shown in the Table the average pore diameter was varied from 0.02 μm to 3.4 μm by changing the process and process parameters. The residual Si in the samples is also listed in Table I. The samples with fine porosity were made by sintering and have no residual Si. In the samples with coarse porosity (made by reaction bonding), residual Si and porosity do not necessarily correlate because of other process variables.

Table I. Properties of porous coupons with fine porosity and coarse porosity

	Fine Porosity				Coarse Porosity			
	SiC-A	SiC-B	SiC-C	SiC-D	SiC-E	SiC-F	SiC-G	SiC-H
Bulk Density (g/cc)	3.01	2.87	2.65	2.36	2.45	2.35	2.33	1.97
Open Porosity (%)	3.0	9.4	16.7	25.7	18.0	22.8	27.0	40.0
Total Porosity (%)	6.1	10.4	17.3	26.3	18.0	25.6	28.9	42.2
Average Pore Diameter (μm)	0.02	0.04	0.06	0.14	3.4	0.56	1.63	1.84
Residual Si (Vol. %)	0	0	0	0	1	5	9	4

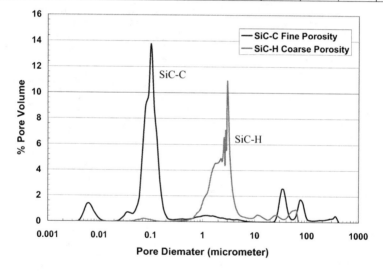

Figure 3. Comparison of porous size distributions of SiCs with fine porosity and coarse porosity.

Detailed microstructural analysis was conducted on all samples to correlate the pore size distribution measured and actual pores in the material. Figure 4 shows such a correlation for the sample SiC-C. Porosimetry on this material shows peaks over the ranges 0.005-0.006 μm, 0.08-

0.2 μm, 1-2 μm, 30-40 μm, 60-100 μm. SEM photos at various magnifications confirm existence of pores in the later four size ranges. The SEM used did not have the resolution to observe pores in the finest range (0.005 to 0.006 μm).

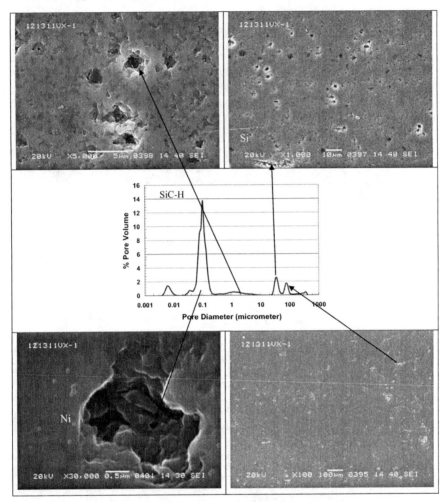

Figure 4. Correlation of microstructural observations at different magnifications with measured pore size distribution for SiC-C (fine porosity).

Figure 5 shows the correlation between porosimetry peaks and actual pores for the sample SiC-H. Porosimetry on this material shows peaks over the ranges 1-4 μm, 10-20 μm, 20-30 μm, and 40-50 μm. Again, SEM photos at various magnifications confirm existence of pores in these size ranges.

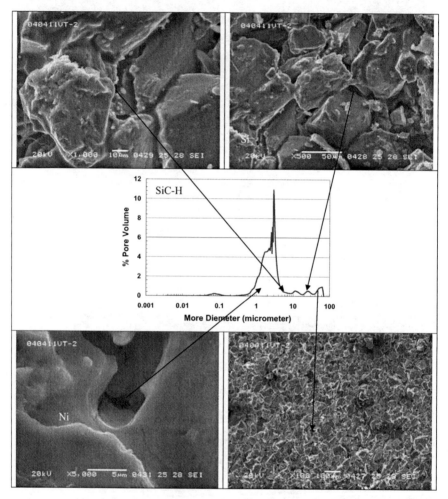

Figure 5. Correlation of microstructural observations at different magnifications with measured pore size distribution for SiC-H (coarse porosity).

The gas permeabilities of the samples are plotted as a function of total measured porosity in Figure 6. Clearly, in addition to the total porosity the pore size also affects the permeability. As expected, the samples with finer porosity, have lower permeability. Overall, samples with permeabilities varying over 5 orders of magnitude have been prepared by changing the process and process parameters. Thus, the permeability can be tailored for specific application.

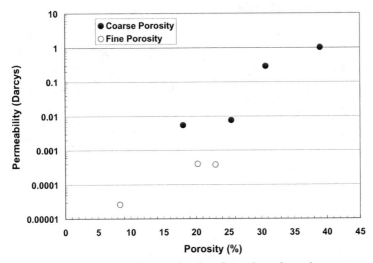

Figure 6. Permeability as a function of pore size and porosity content.

In certain applications (e.g. diesel particulate filters), a combination of micro porosity and macro porosity (channels) is required. Figure 7 shows photos and microstructures of the coupon produced with such a combination of porosity. Figure 8 shows photos of example components.

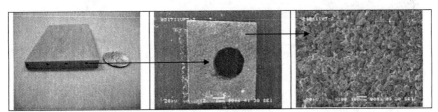

Figure 7. Example of coupon with micro-porosity and macro porosity (channels).

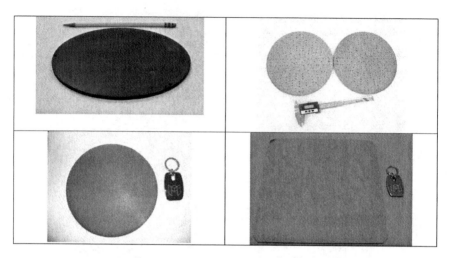

Figure 8. Examples of porous SiC components.

Low and High Purity Porous SiC

For certain porous SiC components, chemical purity is critical to prevent contamination of the use environment as well as the device being placed on the porous component. Table II shows impurity levels in the typical porous components (e.g. 20% porosity, average pore diameter 4 μm). Here, components processed using "standard" raw materials and process environment have impurity levels (parts per million - ppm) as shown in row 2 of Table II. Specifically, B, Al, Ca, and Fe are typically at higher levels. By controlling the raw materials and process environment, porous SiC components with two orders of magnitude lower levels of impurities as shown in row 3 of Table II can be produced. Properties of a selected high purity porous SiC composition are shown in Table III.

Table II. Impurity levels (ppm) in typical standard purity and high purity porous SiC components

Porous SiC Type	B	Al	Ca	Ti	V	Cr	Mn	Fe	Ni	Cu	Zn	Mo
Standard	160	370	280	2.7	13	7	6	250	8	3	5	0.5
High Purity	6.9	68	2	2.9	3.1	0.1	0.01	1.2	0.19	<0.05	<0.05	0.2

Table III. Properties of a selected Si free high purity porous SiC composition

SiC Volume	Porosity Volume	Purity	CTE	CTE	Thermal Conductivity	Electrical Resistivity
%	%	%	ppm/K 20-100°C	ppm/K 20-1350°C	W/m K	ohm-cm
80	20	>99.99	2.6	5.4	120	1

SUMMARY

Due to its high hardness, high temperature capability, excellent chemical resistance, wear resistance, erosion resistance, and high thermal conductivity; porous SiC is a candidate material for a variety of applications such as heat exchangers, diesel particulate filters (DPFs), crude oil filters, membrane filters, burner tubes, sensors, vacuum substrates, catalytic supports etc. In this study, porous SiC was developed with tailorable properties:

- Porosity range 6% to 42%
- Average pore size range: 0.02 to 3.4µm
- Permeability range: 1.0 Darcys to 2.6 x 10^5 Darcys (m² x 10^{-12})
- Purity range: 99.8 to 99.99%
- CTE match with Si wafers

Several porous SiC demonstration components were fabricated. The ability to make components with a combination of micro and macro porosity was also demonstrated.

REFERENCES
[1] Y. Suzuki, "In-situ processing of porous MgTi$_2$O$_5$ ceramics," Ceramic Engineering and Science Proceedings (CESP), Volume 31 [6] (2010) 50-60.

[2] J. Fellows, H. Anderson, J. Cutts, C. Lewinsohn, and M. Wilson, "Strength and permeability of open cell micro-porous silicon carbide as a function of structural morphologies," CESP, Volume 30 [6] (2009) 75-84.

[3] J. Adler, G. Standke, M. Jahn, F. Marschallek, "Cellular ceramics made of silicon carbide for burner technology," CESP, Volume 29 [7] (2008) 10-19.

[4] K. Ohno, "New technology with porous materials: progress in the development of the diesel vehicle business," CESP, Volume 29 [7] (2008) 31-40.

[5] K. Satori, H. Kashimoto, J. Park, H. Jung, Y. Lee, and A. Kohyama, "Thermal insulator of porous SiC/SiC for fusion blanket system," Materials Science and Engineering, 18 (2011) 162019.

[6] V. Suwanmethanond, E. Goo, P. Liu, G. Johnston, M. Sahimi, and T. Tsotsis, "Porous silicon carbide sintered substrates for high temperature membranes," Ind. Eng. Chem. Res., 2000, 39 (9), 3264–3271.

[7] E. Connolly, B. Timmer, H. Pham, J. Groeneweg, P. Sarro, W. Olthuis, and P. French, "An ammonia sensor based on a porous SiC membrane," Solid State Sensors, Actuators, and Microsystems 2005, Volume 2, 1832-1835.

[8] M. Aghajanian, B. Morgan, J. Singh, J. Mears, and R. Wolffe, "A new family of reaction bonded ceramics for armor applications," Proceedings of PAC RIM 4, November 4-8, 2001, Maui. Hawaii, Paper No. PAC6-H-04-2001.

[9] P. Karandikar, S. Wong, G. Evans, and M. Aghajanian, "Microstructural development and phase changes in reaction bonded B$_4$C," CESP Vol. 31 [5] (2010) 251-259.

[10] P. Karandikar, G. Evans, S. Wong, and M. Aghajanian, "Effect of grain size, shape, and second phases on properties of sintered SiC," CESP Vol. 30 [5] (2009) 68-79.

OBTAINING POROUS CORUNDUM CERAMICS BY UTILIZATION OF WASTE RICE HUSK - INVESTIGATION OF COMPOSITION, STRUCTURE AND THERMAL DEGRADATION OF RICE HUSK

Irena Markovska, Bogdan Bogdanov, Dimitar Georgiev and Yancho Hristov

Department of Silicate Tehnology, Assen Zlatarov University, 1 Prof. Yakimov Str., 8010 Bourgas, Bulgaria

ABSTRACT
The aim of the present study is to develop a series of porous ceramic materials providing possibilities to control of their porous structure. Two types of industrial waste were used as initial materials - aluminium oxide from oil refining industry and waste rice husk. The methods of X- ray analysis, differential thermal analysis, scanning electronic microscopy and mercury porosimetery were used mainly. The rice husk are a potential source of silica containing compounds. The white rice husk ash containing more than 90 mass.% SiO_2 and activated rice husk carbon were obtained thermally in this study. The thermal degradation and the structure of rice husk were investigated. The waste $\gamma - Al_2O_3$ is industrial waste from oil industry. It was used as adsorbent or carrier catalyst in several production lines of the petrochemical plant, so it was contaminated with different by type and amount impurities. The contents of impurities decreased after thermal treatment at 1000 ^0C followed by chemical treatment of the product with HCl. Using the recycled alumina and rice husk a series of porous corundum materials were synthesized.

INTRODUCTION
Ceramic filters for liquids excel the other one, e.g. on polymer, vegetable, animal, etc. basis, since they have up to 4 times higher permeability and more than 10 years lifetime.

The open porous corundum ceramic materials (OPCCM) are extremely suitable for the production of filtering elements and systems which are heat and can be used at temperatures above 1000 °C, and pressure up to 20 MPa.[1-3] They are persistent in alkaline and acidic environments, and have long period of exploitation; they can be produced with regulated size of their pores. The open ceramic materials are remarkable with both their great permeability and very good strength properties. Because of these advantages OPCCM are used for purification of waste water, filtration of metals, purification of the different gases from diesel and petrol engines.[4-5] A wide range of component shapes is also available. In addition, it is a simple process to apply a dense coating to one or more surfaces, either to eliminate permeability or to increase the mechanical properties of the surface layers.

For the production of OPCCM, natural and technical materials are used as well as waste materials. The obtaining of OPCCM is possible by the following methods:

1. By extrusion and pressing of samples from ceramics materials – the size of the apertures is dependent of a diameter of the needles of the pressing swage.
1. Method of burning additives – up to 50% porosity.
2. Gas method   - up to 60 or 70% porosity.
3. Direct foaming   - up to 70% porosity.
4. Replication   - up to 90% porosity.
5. Combined method   - establishing a multi layer filters.

The aim of the present paper is to synthesize porous corundum ceramics by the method of burning-out additives. By this method, porosity is obtained as a result of burning of organic or inorganic additives in the blend.[6] The burning-out additives used were waste rise husk.

Rice husk is a by-product of rice milling process and are a major waste product of the agricultural industry. They have now become a great source as a raw biomass material for manufacturing value-added silicon composite products, including ultrafine SiO_2 [7], silicon carbide [8,9], silicon nitride [8,10], fillers of rubber and plastic composites, etc. [11], for preparation of porous ceramics from rice husk and alumina for bioreactor [12], etc. By the addition of raw rice husk as filler in corundum materials, certain amount of silicon oxide remains after their burning which reacts with the aluminium oxide to give mullute phase. Since the coefficient of linear thermal expansion of mullite is lower than that of corundum, its presence improves the thermal stability of the ceramic samples.[13]

In connection with above mentioned the aim of the present work is to obtain porous corundum ceramic material using waste alumina and waste rice husk, which can be used as part of an apparatus for filtering and separation of liquids, emulsions and melts.

EXPERIMENTAL
Methods

The materials obtained were characterized by X-ray analysis, differential thermal analysis (DTA), scanning electron microscopy (SEM) and mercury porosimeters.

The X-ray analyses were carried out by the method of powder diffraction using X-ray apparatus equipped with goniometer URD-6 (Germany) with cobalt anode and K_α emission.

The DTA experiments were performed on an apparatus for complex thermal analysis (STA 449 F3 Jupiter), NETZSCH – Germany.

Porosity of the specimens was determined by an automatic mercury porosimeter Karlo Erba 1520 (Italy), with working pressure up to 150 MPa and opportunities for study of the porous structure of the samples – from 50 Å to 75 000 Å.

The micrographs were taken using scanning electron microscope Tesla BS 340 (Czech Republic). The SiO_2 content in the solid residue was determined after treatment with hydrofluoric acid and the carbon content – by automatic gas analyzer Carlo Erba 1104 (Italy).

Raw materials
In this study waste alumina and waste rice husk were used as raw materials.

The waste γ - Al_2O_3 is industrial waste from Lukoil Neftochim Co., Bourgas. It is used as adsorbent or carrier catalyst in several production lines of the company, so it is contaminated by different type and amount impurities. The aluminium oxide was produced by Alcoa Co. – USA in the form of granules with sizes about 4 mm.

Rice husk (RH) were obtained from suburb areas of Pardjik, Bulgaria. Before use, the rice husk were thoroughly washed – three times with tap water followed by three times with deionised water to remove adhering soil, clay and dust, boiled for an hour to desorbs any impurities and finally, dried at 100 ^{0}C overnight. The dried husk were ground in rotary cutting mill and sieved manually with 0.63–0.12 mm sieves. This starting material was used for all further studies.

Processing of the waste rice husk
The present study was carried out with rice husk obtained during processing of rice variety Krasnodarski 424 grown in Bulgaria. This kind of rice contains 17.8% husk. They are arc-shaped and size approximately: 8 mm length, 2-3 mm width and 0.10÷0.15 mm thickness. The husk contain 74.5% organic matter (cellulose, hemicellulose and lignine) and water, and the rest is inorganic matter containing 20% SiO_2 and 5.5% mixture of the following oxides: CaO, Fe_2O_3, MgO, Al_2O_3, Na_2O, K_2O, MnO_2, as well as traces of Cu and Pb.

Raw rice husk were used for the experiments. They were added to the corundum matrix and the samples were sintered at temperatures higher than 1000°C in nitrogen or oxygen medium. Thus, it became necessary to study the changes taking place in rice husk during their burning in air and nitrogen. The experiments under non-isothermal conditions were carried out in a derivatograph system F.Paulik - I.Paulik - L.Erdey (Hungary). Rice husk samples of 100 mg were used for the experiments. The RH were heated to 1000°C at heating rate of 10°C/min by the following methods:

- In static air. In the course of the heating, first the physically adsorbed water was released (5%), then pyrolysis takes place to form a hard residue (26 %) containing mainly SiO_2.

- Carbonization of raw rice husk in nitrogen medium with flow rate 2.5 $cm^3.min^{-1}$ at 20°C and 1 atm pressure. Again physically adsorbed water was first released and the following pyrolysis of the organic mass gave hard residue (45 %).

Under thermal treatment, the rice husk underwent a series of changes depending on the gaseous medium and treatment time which were established by the derivatographic analysis. Fig.1 (a,b) show the TG curves registered by heating the samples in oxidative and inert media.

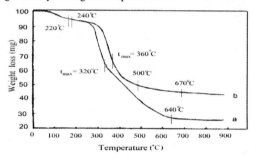

Figure 1. TG curves of: (a) rice husk burnt in air, (b) rice husk burnt in nitrogen medium

The DTA analysis of rice husk treated in oxidative medium (Fig.1 a) showed that the process of thermal destruction began at 220°C and ended at 640°C with 74% mass loss. In nitrogen the thermal degradation began at 240 °C and at 500 °C the weight loss was aleady 50%. At 670°C the process almost ended – the weight loss was 55% (solid residue 45%). Comparing the derivatograms, it can be seen that the processes of destruction proceeded at maximum rate at 320°C in air (a) and at 360°C in nitrogen (b). On the TG curve for the rice husk burnt in air the region with the highest desruction rate was observed in the temperature interval from 220 to 380 °C, while in nitrogen, this region was observed in the temperature interval from 240 to 400°C.

The shape of the TG curves can be explained with the slower oxidative-destruction processes and with the immanent structures of SiO_2 forming husk carcass and the diffusion processes. It is well known that the destruction of silanol groups occurs normally at temperatures of 500-600°C, while in silicagel this process can be observed at 700-800°C. Probably, the products of organic matter destruction facilitate the release of OH - groups from the surface of SiO_2 in rice husk.

Figure 2 shows four kinds of rice husk samples:

A) paddy - rice grain in husk

B) raw rice husk preliminarily washed to remove mechanical impurities and dried at 120°C (RH);

C) rice husk oxidized at 650°C and 1000°C in air – white ash (WRHA);

D) rice husk carbonized at 1000°C in nitrogen – black ash (BRHA).

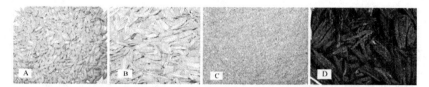

Figure 2. Thermally treated and non treated rice husk: A – paddy; B - RH; C – WRHA and D – BRHA

The analysis of the composition of WRHA showed that, due to the full oxidation of the carbon organic matter at temperatures above 800°C, the solid residue was almost pure SiO_2 (>90 %) with small amounts of other inorganic oxides. Table 1 shows the composition of the oxidized powder.

Table I Composition of WRHA

Component	Quantity, mass. %
SiO_2	94.47
Fe_2O_3	1.32
K_2O	1.08
MgO	1.03
Al_2O_3	0.98
CaO	0.62
MnO_2	0.49
Na_2O	0.01
Cu	traces
Pb	traces

The results from SEM are shown in Figs. 3-5. In the micrographs given further below it can be noticed that the outer relief of the raw and thermally treated husk has the same appearance. This fact shows that SiO_2 is concentrated mainly on the surface of the husk, lending it hardness and water resistance. Scanning electron micrograph of raw husk is presented in Fig. 3.

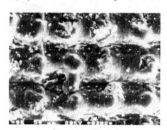

Figure 3. Micrograph of raw rice husk (RH)

In Figures 4 (A, B) micrographs of husk, thermally treated at 1000⁰C are shown: Fig. 4 (A) – after oxidation in an air medium and Fig. 4 (B) – after carbonization in a nitrogen medium.

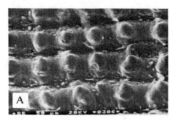

Figure 4. Micrographs of thermally treated rice husk: (A) after burning in air medium, (B) after pyrolysis in nitrogen medium

In both micrographs it can be seen that the husk surface relief consists of the same repeating little "domes". The husk surface did not change after thermal treatment in an air and inert media. It only suffered a certain densification (Figs. 3 4, B) as a result of the releasing of volatile products – CO_2, CO, H_2, H_2O and CH_4. Remarkable structure stability was present in both the raw and thermally treated products. This preservation of the shape is undoubtedly due to the even distribution of SiO_2 on the surface of the starting material, which on its part can be associated with the basic biological role of the husk to preserve the seed from mechanical damage and to retain moisture. The preservation of the shape corroborates the fact that SiO_2 performs the functions of a skeleton, which is clearly confirmed by the micrographs of the thermally treated samples. Repetition of this kind is shown by the objects in the pictures of husk, carbonized in a nitrogen medium (Fig. 5 A) and husk, oxidized in an air medium (Fig. 5 B) at 1000⁰C, whose magnification is four times smaller. In the micrograph in Fig. 5 (B) it can be seen that after the organic mass is separated, mainly the inorganic ingredient, which forms the silicon-oxygen skeleton, remains in the husk particle. From the micrograph shown in Fig. 5 (A) it can be found out that under the silicon-oxygen skeleton a layer of amorphous carbon emerges as a result of the carbonization of the organic part that has taken place.

Figure 5. Micrographs of thermally treated at 1000⁰C husk: (A) carbonized, (B) oxidized

Processing of the waste Al_2O_3

The following scheme for recycling of waste alumina granules was suggested in order to convert γ-Al$_2$O$_3$ into the stable α-form:

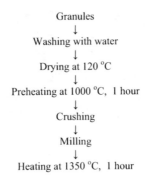

Figure 6. Scheme of recycling of waste γ-Al$_2$O$_3$

The contents of impurities in the initial γ-Al$_2$O$_3$ and the product α-Al$_2$O$_3$ were determined by emission spectral analysis. The results obtained showed that the purity of the product α - Al$_2$O$_3$ was 99.47 % while that of the initial γ -Al$_2$O$_3$ was 99.24 %. The contents of impurities decreased after thermal treatment at 1000°C followed immediately by treatment of the hot product with 4n HCl. The aluminium oxide processed by this technique was the initial material for the synthesis of corundum ceramics.

As a result of above mentioned technological operations, γ- Al$_2$O$_3$ was fully transformed into α-Al$_2$O$_3$. Granulometric composition of the recycled alumina is presented in Table 2.

Table II Granulometric composition of α-Al$_2$O$_3$

fraction, μm	0-5	5-10	10-20	20-30	30-50
Quantity , %	75	8.5	4	4.5	8

Using the recycled Al$_2$O$_3$ (> 99, 47%) and waste rice husk the porous corundum ceramics can be produced with regulated size of its pores.

Synthesis of corundum ceramics by method of burning additions

A series of corundum materials were prepared by the method of burning-out additives from purified aluminium oxide powder containing 5 – 50% ground raw rice husk. The compositions of the materials synthesized are presented in table 3.

Table III Compositions of the samples

№ of composition	α - Al₂O₃, wt. %	RH, wt.%
1	100	0
2	95	5
3	90	10
4	85	15
5	80	20
6	70	30
7	50	50

The specimens were formed by semi-dry pressing on hydraulic press "Carl Zeiss Yena" (Germany) at pressure of 200 MPa.

The samples were sintered in two stages – preliminary and at high temperature. The aim of the preliminary thermal treatment was to burn the plasticizer and was performed in muffle furnace LK-120 (Germany). The preliminary sintering was carried out with intermediate isothermal soaking of 15 min at 100 °C, 30 min at 300 °C, 15 min at 500 °C and monotonous heating to 900 °C where the samples were kept for 60 min. The final high temperature sintering was carried out in superkantal oven Naber (Germany) by heating to 1450 ^{0}C, 1500^{0}C, 1550^{0}C and 1600 ^{0}C in air or nitrogen media.

RESULTS AND DISCUSSION

It is well known that the properties of the materials are strongly affected by their structure and phase composition. The X-ray phase analysis showed that the main phases observed in the samples synthesized in air were corundum, mullite, quartz and cristobalite while for these synthesized in nitrogen – corundum, mullite and cristobalite (fig.7).

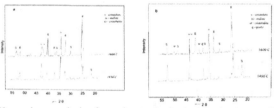

Figure 7. X-ray phase analysis of samples synthesized: a) in nitrogen medium; b) in air

One of the most important methods for increasing the thermal stability is to synthesize ceramics containing phases of low or negative coefficient of linear thermal expansion (CLTE).[6] It is known that corundum belongs to the materials with high CLTE - 8.5 x 10 $^{-6}$ ^{0}C $^{-1}$ (20- 1000 ^{0}C).[14] The presence of phases with low CLTE - cristobalite and quartz (0.5 x 10 $^{-6}$ ^{0}C $^{-1}$) and such an average CLTE - mullite (5.0 x 10 $^{-6}$ ^{0}C $^{-1}$) [14] identified by is prerequisite for the increase of samples thermal stability and it was confirmed by the thermal shock tests carried out.

The thermal stability studies were performed by the method of abrupt change from 1100°C to tap water. The samples of pure corundum were found to crack after less than 10 thermal shocks while the presence of quartz and cristobalite, as well as increased content of mullite, in all the other samples resulted in more than 30 thermal shocks without visible changes.

Study on the porous structure of samples synthesized in air

With respect to the further use of the materials, it is interesting to study the formation of the porous structure in the samples depending on the amount of rice husks added. For this purpose, scanning electron microscopy was used (fig.8 a-c) and mercury porosimetry. The SEM observations showed inhomogeneous porous structure which can be clearly seen in Fig.8a.

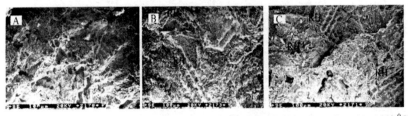

Figure 8. SEM of corundum ceramic sample synthesized in air medium (t = 1550 ^{0}C)

In the micrographs presented, the pores and clear imprints of the burned rice husk built from silicon-oxygen carcass can be seen (Fig.8 b,c). These carcasses are very similar to these shown in the micrographs of the rice husk presented in Figs.3-5. It indicated that during the high temperature sintering, part of the silicon dioxide of the husk reacts with the aliminium oxide matrix to form needle-like mullite crystals (fig.9) while the other part transforms into quartz or cristobalite preserving the general shape of the thermally treated rice husk. As it has been mentioned above, the experiments on thermal stability showed that the increase of the amount of rice husk introduced, respectively SiO_2, results in significant increase of the thermal shock resistance. This increase should be attributed not only to the newly formed mullite phase but also to the specific spring-like structure of the three dimensional structural chains of the silicon-oxygen tetrahedrons which absorb and quench the tensions.

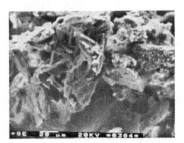

Figure 9. SEM of corundum ceramics

The needle-like mullite crystals obtained as a result of the interaction between the corundum matrix and part of the silicon dioxide contained in the husk can be seen in Fig.9.

Besides, pores different by size and shape can be observed. It was confirmed also by the results obtained from the mercury porosimetry experiments. The data on average radius – R_{av}, pore content in per cent, volume porosity, specific area, average radius on the surface – R_s and average radius by volume of samples containing 10, 30 and 50% rice husk are presented in Table 4.

Table IV Characteristics of the porous structure of samples synthesized in air at 1600°C for 30 min

sample	№ of pore size	R av (Å)	% of pores	Bulk Porousity, cm³/g	Spec. surface, m²/g	R$_{av}$ by surface (Å)	R$_{av}$ by volume (Å)
with 10% RH	1	52500	16.87				
	2	24375	16.87				
	3	13125	16.72				
	4	6094	16.72	0.0678	0.232	5853	17141
	5	3684	16.51				
	6	2155	16.51				
with 30 % RH	1	58929	16.69				
	2	36429	16.69				
	3	24375	33.38	0.0524	0.049	21436	28500
	4	16193	16.66				
	5	10872	16.59				
with 50% RH	1	58929	33.35				
	2	36429	33.35	0.0325	0.019	35118	39918
	3	24375	33.30				

Considering the data in the tables presented, the effect of the amount of rice husk on the formation of the porous structure can be seen. At small amount of the reinforcing component (10%), six types of macrospores were formed uniformly distributed throughout the structure with sizes from 52500 Å to 2155 Å. With the increase of the amount of rice husk introduced, the total number of pores decreased but they had large sizes – from the average radii by surface and volume also increased. It should be noted that all the pores were distributed quite uniformly and were large sized (larger than 1000 Å). It can be seen from the data in table 4 that the porous structure of the materials can be regulated through the amount of rice husk introduced.

Porous structure of the samples synthesized in nitrogen medium

The porous structure of the samples synthesized in nitrogen medium at 1500°C for 1 h. Fig.10 shows a micrograph of the simple containing 20% rice husk.

Figure 10. SEM of sample with 20 % RH, synthesized in nitrogen

In this micrograph can once again be clearly seen the carcass of the carbonized rice husk, as well as the grains of aluminium oxide. The data obtained from mercury porosimetry give an idea about the effect of the rice husk content on materials porosity. The results are presented in Table 5.

Table V Characteristics of the porous structure of samples synthesized in nitrogen at 1550^{0}C for 120 min

sample	№ of pore size	R av (Å)	% of pores	Bulk Porousity, cm^3/g	Spec. surface, m^2/g	R_{av} by surface (Å)	R_{av} by volume (Å)
with 5% RH	1	49038	25.46				
	2	18357	25.46				
	3	10872	25.31	0.0101	0.017	11573	21193
	4	5393	23.77				
with 10% RH	1	49038	7.71				
	2	18357	7.71				
	3	12175	7.73				
	4	9769	7.73				
	5	8162	7.73				
	6	7011	7.73				
	7	6145	7.73	0.0496	0.195	5079	10183
	8	5228	7.70				
	9	4487	7.73				
	10	3908	7.69				
	11	3221	7.68				
	12	2610	7.64				
	13	1919	7.49				
with 15% RH	1	52500	13.36				
	2	24375	20.05				
	3	16193	13.36				
	4	12175	13.36				
	5	9769	6.67	0.0672	0.14	9575	18163
	6	8162	6.67				
	7	6635	6.65				
	8	5228	6.65				
	9	4173	6.63				
	10	3169	6.59				
with 20% RH	1	52500	41.81				
	2	24375	16.71				
	3	16193	8.34				
	4	12175	8.34	0.0631	0.08	15789	30229
	5	9769	8.34				
	6	7673	8.34				
	7	4761	8.34				

Comparing the data in Table 4 and Table 5, it can be seen that much more types of pores were formed by the synthesis in nitrogen medium (Table 5). At low rice husk content uniformly distributed. With the increase of rice husk content to 10 and 15%, the number of 3169 Å, respectively. This great variety of pore types and sizes is most probably due to the influence of the inert medium on the process of corundum matrix sintering. Supposedly, the burning of rice husk is subdued and more gases are released which, in turn, also had their effect on the formation of the porous structure and its diversity. By increasing the percentage of added rice husk again, as in table 4, increased the value of the average radius by surface and volume - from 11 573 and 21 193 Å at 5% to 15.789 and 30.229 Å - at 20% RH, respectively. At the highest rice husk content of 20%, the pore diversity by size and type was lower, so with largest percent are the macrospores with radius 52 500 Å - almost 42%.

Some basic physicomechanical properties of the samples synthesized were determined. The data on density, open porosity and water absorption are presented in Tables 6-9.

Table VI Properties of the samples synthesized in air at $T_{syn} = 1600°C$ for 0,5 h

№ of composition	Quantity of RH,%	water absorption,%	porosity, %	apparent density, $\rho \times 10^{-3}$ kg/m^3
3	10	5.40	15.97	2.95
6	30	14.44	33.92	2.35
7	50	23.33	41.67	1.79

Table VII Properties of the samples synthesized in air at $T_{syn} = 1550$ °C for 1,5 h

№ of composition	Quantity of RH,%	water absorption,%	porosity, %	apparent density, $\rho \times 10^{-3}$ kg/m^3
2	5	7.88	22.26	2.83
3	10	12.53	30.71	2.45
4	15	15.69	34.96	2.23
5	20	18.35	37.77	2.06

Table VIII Properties of the samples synthesized in nitrogen at $T_{syn} = 1450$ ^{0}C for 2 h

№ of composition	Quantity of RH,%	water absorption,%	porosity, %	apparent density, $\rho \times 10^{-3}$ kg/m^3
2	5	6.00	17.80	2.96
3	10	7.07	20.47	2.89
4	15	8.58	23.11	2.69
5	20	17.77	38.41	2.16

Table IX Properties of the samples synthesized in nitrogen at $T_{syn} = 1550$ ^{0}C; 2h

№ of composition	Quantity of RH,%	water absorption,%	porosity, %	apparent density, $\rho \times 10^{-3}$ kg/m^3
2	5	0.52	1.77	3.43
3	10	2.60	8.36	3.22
4	15	3.84	11.48	2.99
5	20	7.77	21.43	2.76

The data in Tables 6-9 indicate that the highest amount of open pores was observed in the samples synthesized at temperature of 1600°C in air for 30 min and at 1450°C in nitrogen where the open porosity was between 41.67% and 37.77% for the samples synthesized in air and between 38.41 and 21.43% for these synthesized in nitrogen. This, in turn, stipulates the higher water uptake and lower density of these compositions. The high porosity in air was due to the high content of rice husk added while that in nitrogen – to the insufficiently high sintering temperature of the corundum matrix. Furthermore, the mullite plays the role of a binder so the samples obtained had comparatively high physicomechanical properties even at such low synthesis temperature.

CONCLUSIONS

Series of porous corundum materials were developed and obtained by the method of burning-out additives by sintering in oxidative or inert medium. Two waste products were used as initial material – waste aluminium oxide and waste rice husk. The burning of rice husk in air gives almost pure SiO_2 – 94.47%. The husk surface did not change after thermal treatment in an air and inert media. It only suffered a certain densification.

A scheme for processing the waste γ – aluminium oxide to obtain α – aluminium oxide was suggested. By this method, samples with high content of open porosity (41.67%) were obtained from the composition prepared from 50% aluminium oxide and 50% rice husk. The X-ray structural analysis showed that the main phases in the samples obtained in nitrogen medium were corundum, mullite, cristobalite, as well as quartz for the samples obtained in air. This is a prerequisite for high thermal stability of the materials, thus making them suitable for use under thermal shock conditions. Microscopic analyses revealed that the husk carcass built of silicon-oxygen tetrahedrons is preserved during the high temperature sintering and it was almost identical ro that of the thermally treated husk.

The porous structure of the samples can be controlled by varying the amount of rice husk introduced and the gaseous medium during the synthesis to obtain materials with predefined properties.

ACKNOWLEDGEMENTS

The financial support of this work by the Bulgarian Ministry of Science and Education under the contract number DDVU-02-106/2010 with the Research Funds Department is gratefully acknowledged.

REFERENCES
[1]B.A. Krasnyi, V.P. Tarasovskii, A.B. Krasnyi, and A.L. Kuteinikova, Effect of filler crystal habit and particle size of a nano- dispersed technological binder of the properties of a porous permeable ceramic material, *Refractories and Industrial Ceramics,* **48** (4), 290-293 (2007)
[2]E.S. Lukin, F.A. Akopov, G.P. Chernyshov, and T.I. Borodina, Strengthened Porous ceramics based on mullite and corundum, *Refractories and Industrial Ceramics* **49** (4), 298-299 (2008)
[3]A. Kirchner, K. MacKenzie, I. Brown, T. Kemmitt, and M.E. Bowden, Structural characterisation of heat-treated anodic alumina membranes prepared using a simplified fabrication process, *J. Membr. Sci.,* **287** (2), 264-270 (2007)
[4]S.A. Polyakov, and Z.I. Sakharova, Ceramic filters for potable water purification, *Glass and ceramics* (7), 17-19 (1997).
[5]Y.-F., Liu, X.-Q. Liu, G. Li, and G- Y., Meng, Low cost porous mullite-corundum ceramics by gelcasting, *J. Materials Sci.*, **36** (15) 3687-3692 (2001)
[6]Technology of ceramics products, edited by S. Bachvarov, *Technics*, Sofia, 2003.
[7]N. Varghese, K. Vinod, Shipra, A.Sundaresan, and U. Syamaprasad, Burned rice husk: An effective additive for enhancing the electromagnetic properties of MgB_2 superconductor, *J. Am. Ceram. Soc.,* **93** (3), 732-736 (2010)
[8]I. G. Markovska, L. A. Lyubchev, G. H. Davarska, X - ray studies of phase transformations in the system rice husk - fibres with high SiO_2 content, *International Ceramic Review (Interceram),* **47** (5), 318 – 321 (1998)
[9]N. Kavitha, M. Balasubramanian, and Deval Vashistha, Optimization of processing conditions on the yield of nano SiC powder from rice husk, *Adv. Materials Research*, **341-342**, 103-107 (2012)
[10]V. Pavarajarn, R. Precharyutasin, and P. Praserthdam, Synthesis of silicon nitride fibers by the carbothermal reduction and nitridation of rice husk ash, *J. Am. Ceram. Soc.*, **93** (4), 973-979 (2010)
[11]P.N.B.Reis, J.A.M. Ferreira, and P.A.A. Silva, Mechanical behaviour of composites filled by agro - waste materials, *Fibers and Polymers*, **12** (2), 240- 246 (2011)
[12]T. Torikai, T. Ishibashi, K. Egoshi, M. Yada, and T. Watari, T., *Key Engineering Materials* **247**, 433-436 (2003)
[13]I. G. Markovska, B. I. Bogdanov, L. A. Lyubchev, I.G. Chomakov, and J. H. Hristov, Thermally stable composites based on waste products, *Tile & Brick International*, **19** (3), 158 -161 (2003)
[14]Materials Science and Technology, v. 11, Structure and properties of Ceramics, edited by R.W.Cahn, P. Haasen, E.J. Kramer, VCH Verlagsgesellschaft Weinheim, Germany 1994.

PROCESSING, MICROSTRUCTURE AND PROPERTIES OF RETICULATED VITREOUS CARBON FOAM MANUFACTURED VIA THE SPONGE REPLICATION TECHNIQUE

David Haack and Rudolph Olson III
SELEE Corporation
Hendersonville, NC, 28792, USA
www.selee.com

ABSTRACT

Reticulated vitreous carbon foam manufactured via the sponge replication technique has a much more homogeneous, more open structure relative to traditional ceramic foam manufactured by similar methods. To characterize this difference, the flow properties of various carbon foam samples having different pore sizes and densities were measured and compared to those for conventional ceramic foam. The part-to-part variation for carbon foam proved to be much lower than that for ceramic, and equivalent pressure drops were achieved in carbon foam at much smaller pore sizes. These and other aspects of the processing, microstructure and properties of carbon foam will be discussed.

INTRODUCTION

Reticulated vitreous carbon (RVC) foam can be manufactured in a variety of ways. For a broad review of the subject, see Klett.[1] This paper deals specifically with RVC foam produced via the sponge replication technique. The basics of the process are very similar to that used to make ceramic foam as first defined by Schwartzwalder[2]; reticulated polyurethane foam is used as the precursor material in both processes, but in the RVC process, the polyurethane is infused with a high carbon bearing resin instead of being coated with ceramic slurry. The excess resin is expressed from the precursor foam, then cured to crosslink the polymer and fired in a reducing atmosphere to produce vitreous carbon. The first patent detailing the preparation of RVC foam via sponge replication was issued to Moutaud, et. al. in 1969.[3] Variants on this patent soon followed, which include production of anisotropic foam[4] and activation of the foam surface.[5] Now, continuous production methods are capable of producing large blocks of RVC foam on the order of 15x20x30 cm. Critical processing parameters to make good foam include polyurethane foam type and density, resin type and properties, thermal profile, and cover gas type and flow rate

RVC foams have unique properties that engender their use to a variety of applications, including fluid flow and renewable energy applications (e.g. air filtration, water filtration, catalyst supports, fuel cells, batteries and advanced capacitors). The unique reticulated structure is highly permeable and contains a high geometric surface area. Figure 1 shows the geometric structure of RVC foam having a cell size of approximately 80 pores per linear inch (ppi). The material is formed of the ligaments of dodecahedral cells. Openings in the cells (called windows) admit thru-flowing fluids, which are partially diverted by the cell ligaments. The foam structure is excellent for encouraging small-scale fluid mixing and dispersion. Bulk densities near 3% of theoretical are easily achieved.

The structure of RVC is much more open than that of traditional ceramic foam. Figure 2 shows the geometric structure of a typical 50-ppi reticulated ceramic foam produced using traditional sponge replication, where the density is more than three times that of the carbon foam in Figure 1. It is apparent that the carbon foam is a truer representation of the original reticulated polyurethane foam. This results from the fact that polyurethane absorbs resin, whereas ceramic slurry can only coat the surface of polyurethane and thus generates a shell. The struts in carbon foam are solid, whereas they are hollow in the ceramic foam, as shown by the polished cross-sections of Figures 3 and 4 respectively. In addition, preventing pore blockages in relatively fine ceramic foam (30-80 pores ppi) is difficult due to the inherently high viscosity and surface tension of aqueous-based ceramic slurry,

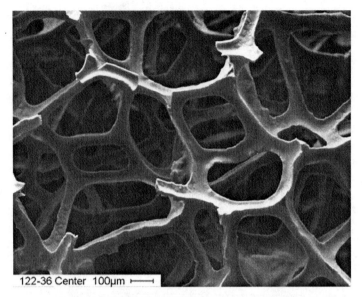

Figure 1. SEM image of RVC foam structure (80-ppi)

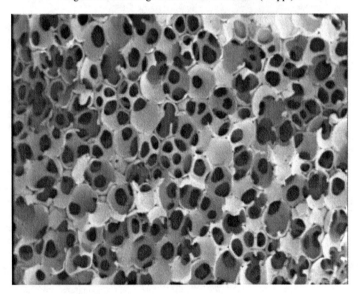

Figure 2. Image of ceramic foam structure (50-ppi, Field of view = 12-mm).

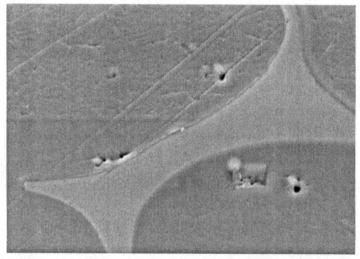

Figure 3. Polished cross-section of RVC foam. Note solid struts of the RCV resulting from the manufacturing methods used. Field of view = 500 microns.

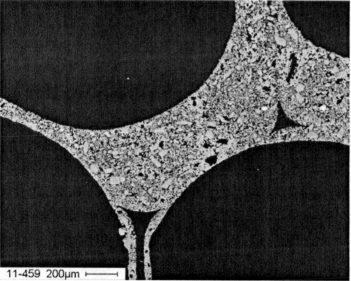

11-459 200µm ├─────┤

Figure 4. Image of a polished cross-section of ceramic foam. Note the triangular shaped voids in locations originally filled by polyurethane.

as well as the hydrophobic nature of polyurethane foam. RVC can be produced at relatively fine ppi values with virtually no pore blockages. Thus, for certain select applications, RVC foam can be a much more efficient structure than traditional ceramic foam, and this is demonstrated here using water permeability measurements comparing RVC foam to porous ceramic foam structures. Other properties such as strength and electrical resistivity are discussed.

EXPERIMENTAL

Ceramic and carbon foam cylinders roughly 40mm in diameter were tested for water permeability over a flow range from 10-40 L/min. The test apparatus is shown schematically in Figure 5. In the test, water is circulated from a 100 gal storage tank through a Brooks model MAG 7400 magnetic flow meter (accuracy +/- 0.1 L/min), then through the test sample. Pressure loss across the sample is measured with an Ashcroft GC52 differential pressure transmitter (accuracy +/- 0.5% of reading). Pressure taps were located 8 diameters upstream and 5 diameters downstream of the test section. The test section is positioned horizontally to minimize the water head effect on the pressure drop measurements. Test samples were cored from foam blanks. Carbon foam samples were manufactured to pore sizes of 20, 40 and 60 ppi and porosities between 95-97%.

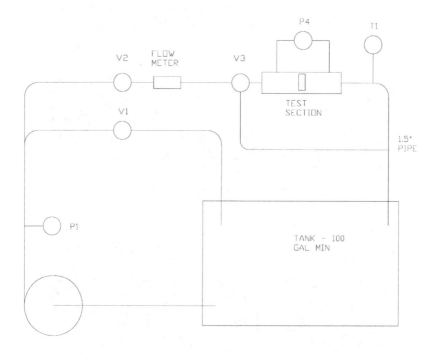

Figure 5. Schematic of test apparatus used to measure water permeability.

RESULTS AND DISCUSSION

Pressure Drop

For product design purposes, it is greatly helpful if the pressure drop information can be fit to a mathematical model so that end-users may easily incorporate carbon foam materials into their system designs. To generate this model, it was assumed that the carbon foam pressure drop (dP/dx) may be accurately modeled through the use of the Hazen-Dupuit-Darcy equation[6]:

$$\frac{dP}{dx} = \frac{\mu}{K_1}U + \frac{\rho}{K_2}U^2 \qquad (1)$$

where ρ is the density of the fluid, μ is the viscosity of the fluid (water in this case), U is velocity – taken as the approach velocity – and K_1 and K_2 are the viscous and inertial resistance coefficients, respectively. The units for K_1 and K_2 are m^2 and m, respectively.

The pressure drop over a free stream velocity range of 0.2 – 1.5 m/s was collected. The data are shown in Figure 6 as a plot of non-dimensional pressure versus the PPI-based Reynolds number. The Reynolds number in the plot is defined as follows:

$$\text{Re}_{ppi} = \frac{\rho U}{\mu PPI} \qquad (2)$$

where PPI is the foam pore size designator expressed as pores per linear inch (units in^{-1}).

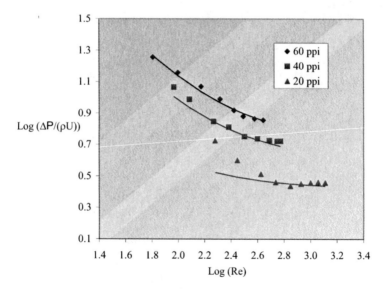

Figure 6. Carbon foam permeability data relating the dimensionless pressure drop against the Reynolds number based on carbon foam pore size as defined by the PPI product characteristic.

A best fit model was generated from these data relating K_1 and K_2 to carbon foam pore size and density, yielding the following equations:

$$K_1 = 9.91x10^{-4}(PPI)^{-2.9} \tag{3}$$

$$K_2 = 0.059 - 0.027(PPI)^{0.047} - 0.042(\rho_{rel})^{0.17} \tag{4}$$

where ρ_{rel} is the sample density relative to the theoretical density of vitreous carbon, presumed to be 1.55 g/cc. These relations, as used in the model of equation (1), yield a reasonable fit to the experimental data shown by the solid lines in Figure 6, provided that the PPI is accurately measured for the specific sample of interest, and that the relative density of the carbon foam is known for the sample. Standard pore size samples were used to estimate the pore size of each individual sample.

In this study, no attempt was made to relate the permeability coefficients to physical characteristics of the carbon foam (e.g. pore, window size or ligament diameter). Instead, these relations were generated to serve as a design tool for customers looking to use SELEE carbon foam in through-flow applications, such as heat exchange, fluid filtration or catalyst substrates. It is interesting to compare how the carbon foam permeability differs from conventional ceramic foams used in similar applications under similar conditions.

In the ceramic foam case, the structure of the foam differs considerably from that of the carbon materials for two distinct reasons: 1) The ceramic foams are made necessarily more dense than the carbon foams to achieve reasonable strength in the ceramic product, and 2) the physical characteristics of the ceramic foam are considerably different due to how the foam is made through the foam replication process. Figure 7 shows the pressure drop measured for typical samples of ceramic and carbon foams of roughly similar pore size designation (15-ppi for ceramic foam and slightly finer 20-ppi for carbon foam). The pressure drop penalty paid by the ceramic foam results from a significantly reduced porosity (84% as compared to 97% for the carbon foam), and a less "clean" pore structure resulting from blockages (as demonstrated in Figure 2). The necessarily less clean structure of the ceramic foam results from the relatively high density required satisfying end use strength requirements. This problem is significantly reduced with carbon foam, as its rupture strength tends to be superior to that of ceramic foam at higher porosity.

Study of the Pressure Loss Variation

Variation in product pressure drop due to normal manufacturing variances is an important consideration to end users of the product. Variation in density and pore size will directly influence the permeability of the product, causing unwanted swings in pressure losses in use. To quantify the extent to which pressure loss variation is expected, the relations for K_1 and K_2 were studied for the maximum expected density and pore size variation in the manufacturing process.

Carbon foam relative density variation results from swings in processing related to pore size, dimensional size and accuracy of application of carbon forming materials. Production measurements show the standard deviations of density variation amount to a maximum variation of approximately 30% (i.e., relative density ranging from about 2.5% to 3.5% for a typical product). Carbon foam pore size variation results from swings in pore size related to physical, environmental and raw material variations in the foam manufacturing process. Generally, the product is divided into foam "grades" specified by pore size. For pore size ranges near 20 ppi, these swings in pore size amount to about 25% of the nominal pore size.

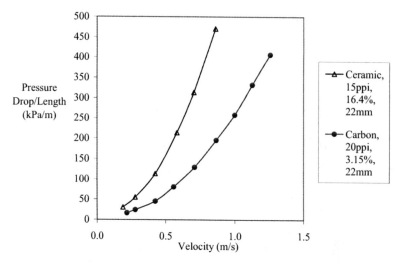

Figure 7. Pressure drop comparison for ceramic and carbon foam having densities of 16.4 and 3.15%, respectively, and both with 22-mm thickness. Carbon foam shows much lower pressure drop despite being fabricated at a smaller overall pore size.

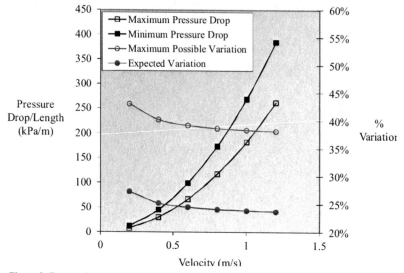

Figure 8. Expected and maximum pressure drop variation versus velocity for carbon foam.

Using the K_1 and K_2 relations as evaluated with the anticipated variable ranges for relative density and pore size designation, and applied to equation (1) yields the expected variation within a carbon foam standard product. The maximum possible variation between the highest and lowest pressure drop within a product range is shown in Figure 8. Variation in pressure drop can be expected to be as high 40% due to normal variability in the product density and pore size. It should be noted that this degree of variability will be rare (where both the maximum level of variation exists for both density and pore size in the same part). A more reasonable estimation would be to evaluate the variation at the +/- 2 standard deviation level, where about 25% pressure drop variation from piece to piece within a production lot is expected to exist.

In an attempt to compare the level of pressure drop variation induced by the carbon foam to conventionally manufactured ceramic foams, a similar exercise was performed. Permeability coefficients K_1 and K_2 were developed for ceramic foams in standard product pore size ranges. These coefficients were used to develop a model that is in-turn used to calculate the maximum and minimum pressure drop of the foam subject to water flow at a particular target flow velocity, given expected ranges of variation for product density and pore size. For similar pore size ceramic foam with target density range between 10-14% of theoretical density, the maximum expected variation in pressure drop is about 62%, with a typical variation of about 35%. This degree of variation is much higher than that expected to be seen with the carbon foam. The improved structure and higher porosity contribute to lesser levels of variation for the product as used in flow-intensive environments.

Strength and Compressed RVC Foam Structures

A comparison of the modulus of rupture (MOR) between RVC foam and ceramic foam (phosphate glass-bonded alumina) is shown in Figure 9 for typical relative density ranges for each product. Because of the near-ideal ligament formation for the carbon foam material, the strengths tend to be much higher than those achieved with ceramic materials. Because of the low weight of the carbon foams materials, their specific strength (MOR divided by sample density) is greater than that of typical ceramic foams by more than an order of magnitude, opening the door to applications where lightweight, strong and highly porous materials are desired.

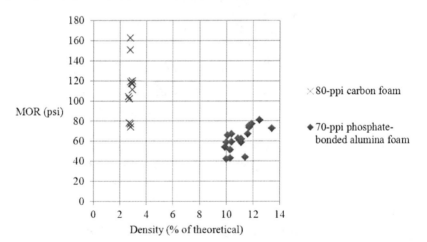

Figure 9. Modulus of rupture versus density for carbon and phosphate glass-bonded alumina foams.

Some applications require strengths greater than can be achieved with the RVC foam described above. One way to enhance its strength is to compress or "felt" the polyurethane foam precursor prior to resin infiltration and curing, as has been demonstrated in the patent literature[3]. Figure 10 displays a compressed RVC foam structure where the compression of about 10:1 in one direction was applied to the polyurethane foam prior to curing. This process increases product density and thereby strength without introducing pore blockages in the structure due to excess vitreous carbon. This would be impossible to achieve using conventional ceramic foam processing methods. The crush strength is shown in Figure 11 for samples prepared at different porosity levels. Obviously, increasing product compression leads to increasing product density and strength.

Control of cell compression also enables the control of other product properties and characteristics, including pore size, permeability, and electrical and thermal conductivity. Figure 12 shows the electrical resistivity of compressed carbon foam materials over a range of final product porosity, as measured using a standard 4-point probe measurement technique. Resistivity is found to increase sharply as the total porosity of the material increases. The trend in decreasing resistivity is expected to continue at lower levels of product porosity, eventually folding over to approach the dense amorphous carbon resistivity value of about 0.004 Ω-cm. Creative control of properties can help end users to optimize their designs by allowing them to tailor the carbon foam component to suit the function of the design more exactly.

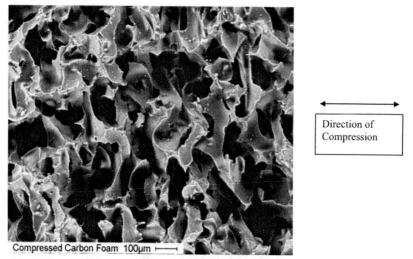

Direction of Compression

Compressed Carbon Foam 100µm ⊢────┤

Figure 10. SEM image of compressed cell carbon foam. Cell compression serves to increase the density the material without generating blocked pores, significantly enhancing strength.

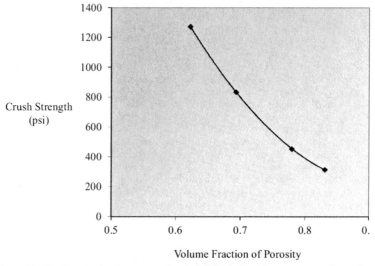

Figure 11. Crush strength of compressed carbon foam prepared over a range of porosity.

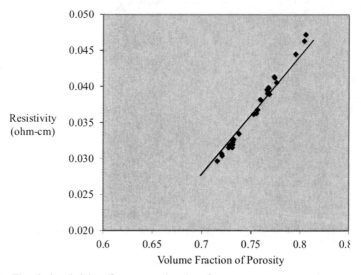

Figure 12. Electrical resistivity of compressed carbon foam structures measured over a range of porosity. Resistivity significantly decreases as porosity decreases for the materials studied.

CONCLUSIONS

In catalyst and filtration applications, the use of carbon foam shows the potential to significantly enhance performance over ceramic foam materials because a finer overall pore size can be used with a lower resulting pressure drop. These performance enhancements result from the excellent foam structure replication made possible using resin-infiltration in conjunction with the sponge replication manufacturing technique. Cells of RVC material are fully clean with little to no observable blocked pores within the structure. This enables the product to be not only highly porous, but display a very high specific strength, more than 10x greater strength per unit density than typical ceramic foams.

Further, the material can be made in a compressed state, where the product is compressed in one or more directions prior to resin curing. The resulting moderate porosity material has significantly elevated strength and easily tailored physical properties. It is possible to control material properties such as strength, conductivity, and permeability accurately with this method of product manufacture, offering product designers additional flexibility to optimize their designs.

REFERENCES

[1]J. Klett, Carbon Foams, *Cellular Ceramics*, ed. M. Scheffler and P. Colombo, Wiley-VCH, 137-157, 2005.
[2]K. Schwartzwalder and A. V. Somers, Method of Making Porous Ceramic Articles, US Patent 3,090,094, May 21, 1963.
[3]G. Moutaud, Neuilly-sur-Seine, Hauts-de-Seine, Process for the Manufacture of Macroporous Vitreous Carbon, US Patent 3,446,593, May 27, 1969.
[4]C. H. Franklin, C. S. Vinton, and H. C. Geen, Reticulated Anisotropic Porous Vitreous Carbon, US Patent 4,067,956, Jan 10, 1978.
[5]C. S. Vinton and C. H. Franklin, Activated Reticulated or Unreticulated Carbon Structures, US Patent 4,154,704, May 15, 1979.
[6]V. Loya, M. Medraj, E. Baril, L.P. Lefebvre, and M. Gauthier, "Permeability of metallic foams and its dependence on microstructure," Light Metals 2005, 44th Annual Conference of Metallurgists of CIM, Calgary, Alberta, Canada, 2005.

AIR-ATMOSPHERE SINTERING OF Si$_3$N$_4$-BASED POROUS AND FOAMED CERAMICS

Philip T.J. Gagnon, Amit K. Gandhi and Kevin P. Plucknett[*]
Dalhousie University, Materials Engineering Program, Department of Process Engineering and Applied Sciences, 1360 Barrington Street, PO Box 15000 Halifax, Nova Scotia, B3H 4R2, Canada.

ABSTRACT

Porous and foamed ceramics, based primarily on Si$_3$N$_4$, have been developed using a simple air-atmosphere sintering procedure. This approach entails enveloping the samples being sintered within a protective Si$_3$N$_4$ powder bed, such that subsequent heating in an air atmosphere furnace results in oxidation of the outer regions of the powder bed. The outer SiO$_2$ scale that is formed then protects the enclosed Si$_3$N$_4$ samples from subsequent oxidation. In the present work two families of material have been developed. The first is based on porous β-Si$_3$N$_4$ ceramics prepared through partial sintering, in this instance using single, rare-earth sintering additions to minimize densification. The second group of ceramics is based on a novel Si$_3$N$_4$-Si$_2$N$_2$O-Bioglass® composite, which foams during the sintering heat-treatment. The effects of the sintering treatment on each material are highlighted, from the perspectives of microstructure development and phase transformation behavior.

INTRODUCTION

Ceramics based on silicon nitride (Si$_3$N$_4$) are used in a wide variety of engineering applications due to a favorable combination of mechanical properties and environmental durability (e.g. oxidation and corrosion resistance).[1,2] One of the major limitations to wider implementation of Si$_3$N$_4$ based ceramics is related to the processing costs of the materials, which arises from both the raw materials and the fabrication expenses. In terms of preparing Si$_3$N$_4$-based ceramics, it is conventionally viewed that an inert atmosphere is necessary, notably nitrogen, such that dissociation and oxidation phenomena are avoided. However, it has recently been demonstrated that dense β-Si$_3$N$_4$ ceramics can be prepared using a low-cost air atmosphere furnace,[3,4] with the materials being held within a pure Si$_3$N$_4$ powder bed during sintering. When heating in air, the outer, exposed surface of the powder bed oxidizes, forming a protective, continuous silicon dioxide (SiO$_2$) scale. As a consequence, the samples held within the interior of the powder bed are protected from the oxidizing atmosphere and can be sintered to near-theoretical density without degradation.[4]

The use of this air-atmosphere sintering approach has also recently been extended to prepare porous Si$_3$N$_4$ ceramics.[5,6] The retention of porosity in the final product entails 'partial sintering' of the material, for example through the use of a single, refractory oxide addition such as a rare-earth oxide. As a consequence, the α- to β-Si$_3$N$_4$ transformation is promoted, while densification is retarded and a reasonable level of porosity can then be retained (typically 20 to 40 % by volume, depending upon the processing route and sintering additives employed). In this instance it can be seen that further processing complications arise due to the necessary presence of retained porosity in the materials throughout the sintering cycle, with the consequent potential for increased degradation (e.g. through internal oxidation). In the present work, recent progress in the air-atmosphere sintering of porous β-Si$_3$N$_4$ is described, noting in particular the effects of processing parameters such as heating rate, pore volume (varied through the use of fugitive carbon fillers), and the rare-earth oxide sintering additive used. In addition, the development of a novel Si$_3$N$_4$-based foamed composite is outlined, based on a final component mixture of Si$_3$N$_4$, silicon oxynitride (Si$_2$N$_2$O) and Bioglass® (equivalent to the commercial grade 45S5 glass). This secondary study highlights aspects relating to the microstructural development of the foams and their phase evolution during the sintering heat-treatment.

---

[*] Corresponding Author (email: kevin.plucknett@dal.ca)

EXPERIMENTAL PROCEDURES

Preparation of Porous Si_3N_4

Porous Si_3N_4 ceramics have been prepared using Ube SN E-10 α-Si_3N_4 powder, together with a variety of micron-sized, single rare-earth oxide additions (summarized in Table 1); all porous materials were prepared with an equivalent additive content of 3.2 mol. % rare-earth oxide. To vary the retained pore content, graphite filler (with a nominal particle size of 8-10 μm) was also incorporated into selected samples. The graphite content was varied up to ~13 wt. %. Powder batches were wet ball-milled for 24 h in propan-2-ol, dried and then sieved through a 75 μm stainless steel sieve. Pellets were then uniaxially compacted at ~30 MPa, followed by cold-isostatic pressing at ~170 MPa. After compaction, samples with graphite filler were heat treated in air at 650°C for 24 h to ensure complete carbon removal, following prior studies.[7] Sintering was performed in various air-atmosphere sintering furnaces, with the samples supported in the same α-Si_3N_4 powder within aluminum oxide (Al_2O_3) crucibles; for selected compositions a 'sacrificial' pure α-Si_3N_4 pellet was placed near the top of the powder bed, above the actual samples being sintered, as this was observed to eliminate the potential for the samples to be partially uncovered during sintering. Further details regarding this sintering arrangement can be found in the prior publications.[4-6] Two differing sintering cycles were used for the present work (as shown in Figure 1(a)), which are termed 'slow' and 'fast' heating in the subsequent text. These heating rates were the maximum available for the furnaces used during specific stages of the present study. Sintering temperatures up to 1750°C, held for 2 h, were subsequently applied for the preparation of porous β-Si_3N_4 ceramics.

Table 1. Rare-earth oxide sintering additives used for porous Si_3N_4 preparation.

Additive	Supplier	Purity (at. %)
Gd_2O_3	Treibacher Industries, Toronto, Canada	99.99
La_2O_3	Metall Rare Earth, Shenzhen, China	99.99
Nd_2O_3	Treibacher Industries, Toronto, Canada	99.99
Sc_2O_3	Treibacher Industries, Toronto, Canada	99.9
Sm_2O_3	Treibacher Industries, Toronto, Canada	99.99
Y_2O_3	Metall Rare Earth, Shenzhen, China	99.99
Yb_2O_3	Treibacher Industries, Toronto, Canada	99.99

Preparation of Si_3N_4-Si_2N_2O-Bioglass® Composite Foams

The production of Si_3N_4-Si_2N_2O-Bioglass® composite foams followed the same general experimental approach as used for the porous β-Si_3N_4 ceramics, in terms of the α-Si_3N_4 packing powder and sample crucible arrangement, with a 'sacrificial' pure α-Si_3N_4 pellet sited above the actual sample during sintering. The same grade of Ube α-Si_3N_4 powder was employed for foam production (for both samples and the packing bed). The sintering additive constituents were designed to be of the same composition as the Bioglass® grade 45S5 (nominal constituents in wt. %: SiO_2 (45 %), CaO (24.5 %), Na_2O (24.5 %), P_2O_5 (6 %)), and were prepared in an initial form as a mixture of sodium and calcium carbonate, along with sodium phosphate hexahydrate. The decomposition behavior of each of these individual components was assessed using thermogravimetric analysis (TGA), which allowed the subsequent tailoring of the heat-treatment profile to allow decomposition without damaging the samples. Following the preliminary TGA studies, this was determined to be ~800°C.

In addition, based on an initial aim of reaction forming Si_2N_2O in the final product (i.e. a Si_2N_2O/Bioglass® foamed composite), from a 1:1 molar ratio mixture of α-Si_3N_4 and SiO_2, the addition of an appropriate amount of additional fine SiO_2 powder was also made, on top of that required for Bioglass® formation. Pellets were prepared from the raw constituents following the same method as the porous Si₃N₄ ceramics. To allow for decomposition of the carbonate and hydrate containing components, the sintering cycle used a lower initial heating rate, up to the nominal onset temperature for oxidation of Si₃N₄ (i.e. 800°C), and then a short hold at that temperature, followed by a high rate ramp as before. The general heating cycle used for the foamed ceramics is shown schematically in Figure 1(b). In order to assess the sintering/foaming response, the Si₃N₄-Si₂N₂O-Bioglass® composites were prepared at temperatures between 1300 and 1675°C, with a hold period of 2 h at the final heat-treatment temperature. Varied sintering times were also explored at the highest temperature, with samples held for between 1 and 4 h at 1675°C.

Characterization Procedures

The densities of the porous Si₃N₄ ceramics were determined via immersion in mercury, due to the retained open porosity. Conversely, the densities of the closed cell Si₃N₄-Si₂N₂O-Bioglass® foams could be measured in distilled water. Dimensional changes of the foams after sintering were determined using both digital calipers and a micrometer. The microstructure of fracture surfaces for both of the air-atmosphere sintered families of material were assessed using field emission scanning electron microscopy (FE-SEM), with all of the samples lightly carbon evaporation coated to prevent charging. Crystalline phase analysis was conducted using x-ray diffraction (XRD), with subsequent phase identification achieved using the ICDD crystallographic database.

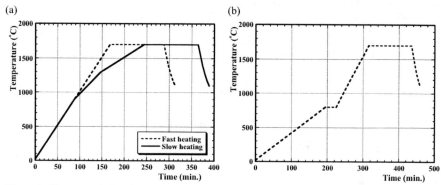

(a)   (b)

Figure 1. The general sintering cycles used for: (a) porous Si₃N₄ ceramics and (b) Si₃N₄-Si₂N₂O-Bioglass® composites. In each case a nominal maximum temperature of 1700°C is highlighted for illustrative purposes only.

RESULTS AND DISCUSSION

Porous Si₃N₄ Ceramics

Initial studies relating to the preparation of porous Si₃N₄ ceramics in an air-atmosphere furnace highlighted the formation of a graded Si₂N₂O-rich surface region, due to partial oxidation of the sample inside of the protective powder bed (with an associated sample weight gain); in this instance the 'slow' heating profile from Figure 1(a) was applied.[5] Aside from the formation of Si₂N₂O, another notable aspect of sintering porous Si₃N₄ in air using the 'slow' heating rate was that the final densities were notably higher than similar compositions prepared in a nitrogen environment, which can be attributed to viscous-phase sintering being promoted by the excess SiO₂ formed through sample oxidation.[5] Subsequent evaluation of the 'fast' heating rate (Figure 1(a)) has shown that the formation of Si₂N₂O can be fully eliminated and lower sintered densities arise, as demonstrated in Figure 2; in this instance

5 wt. % yttrium oxide (Y₂O₃) addition was used (~3.2 mol. %). Under these conditions oxidation of the surface of the powder bed occurs sufficiently rapidly to then prevent the subsequent *internal* oxidation of both the enveloping protective α-Si₃N₄ powder bed and the sample sited within it. In this instance it can be seen that the sintered surface is then comprised essentially entirely of the desired β-Si₃N₄ phase, while the final sintered density is reduced by close to 10 % (as viscous phase sintering can be eliminated). This change is attributed to the avoidance of sample oxidation when using the faster heating rate. At the slower heating rate the oxidation product that forms within the sample, SiO₂, allows some degree of viscous phase densification to occur prior to reaction with Si₃N₄ to form Si₂N₂O, giving rise to the two issues highlighted in Figure 2.

Having established the benefits of utilizing the 'fast' heating schedule, all of the following data outlines studies conducted using this higher heating rate for sample preparation. It is likely that the benefit from further increasing the heating rate would be limited, and would start to increase the probability of crucible failure due to thermal shock. In terms of modifying the extent of porosity, fine graphite filler particles have been incorporated into the green-body formation stage, which are then removed to leave a more open Si₃N₄ preform. Confirmation of the complete removal of the filler particles is shown in Figure 3(a), where the weight loss after heat-treatment at 650°C in air corresponds closely to the initial graphite content. The effects of the initial graphite filler content upon the final sintered density are demonstrated in Figure 3(b). It is apparent that the extent of retained porosity can be further extended through incorporation of the fugitive filler, with a further decrease in density of ~10 % relative to the samples prepared without filler. It should be noted that in the present work samples were isostatically pressed at 200 MPa. However, it is expected that a further reduction in density could be achieved through the elimination of this second compaction stage, and simply uniaxially pressing at a lower pressure.

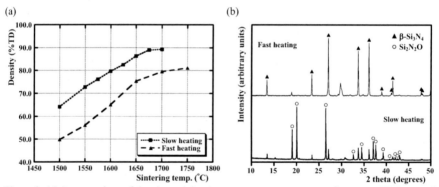

Figure 2. (a) A comparison of the sintered density vs. temperature response for porous Si₃N₄ ceramics (with 3.2 mol. % Y₂O₃ additions) prepared using both 'slow' and 'fast' heating rate cycles. (b) The surface phase composition of porous Si₃N₄ sintered at 1700°C for 2 h using both the 'slow' and 'fast' heating cycles, highlighting the elimination of surface Si₂N₂O formation for the latter case.

In addition to assessing the effects of fugitive fillers, the influence of the rare-earth oxide used has also been assessed. In this instance a selection of lanthanide and Group III oxides have been initially evaluated, using a constant oxide content of 3.2 mol. % (equivalent to that used with the earlier Y₂O₃ additions). The effects of rare-earth oxide upon the densification behavior of air atmosphere sintered Si₃N₄ ceramics are presented in Figure 4(a). It is apparent that relatively high sintered densities are again obtained, with small variations from one oxide to another, typically higher than when using Y_2O_3 additions. In comparison to samples sintered in nitrogen, which typically leads to densities of the order of 60-70 % of theoretical,[8] sintering in air can result in densities in excess of 90 % of theoretical. The reason for this is unclear at the present time, and will be studied in more depth in the future. However, one clear aspect relating to the differences in nitrogen and air processing are that weight losses are typical when sintering in nitrogen (i.e. at 0.1 MPa), which may gibe rise to the lower sintered densities as Si_3N_4 decompositions will result in an internal pore pressure that may counteract densification. Evaluation of the α- to β-Si_3N_4 phase transformation behavior highlights relatively small differences across the various rare-earth oxides, as shown in Figure 4(b), which is again contrary to observations made when sintering similar compositions in nitrogen.[8] The reasons for these differences in sintering and phase transformation response are not clear at the present time.

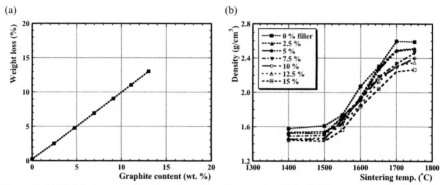

Figure 3. (a) Weight loss as a function of graphite filler content following heat-treatment at 650°C for 24 h. (b) The effects of sintering temperature upon density for samples prepared with varying initial graphite filler contents (note that the marked graphite content is the amount added *in addition* to the initial 100 wt. % Si_3N_4 plus Y_2O_3 mixture).

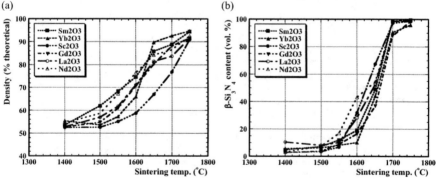

Figure 4. The effects of sintering temperature on porous Si_3N_4 ceramics prepared with various rare-earth oxide additions (at a constant level of 3.2 mol. %): (a) densification behavior and (b) the α- to β-Si_3N_4 transformation response.

Si_3N_4-Si_2N_2O-Bioglass® Foams

The development of Bioglass® containing foamed ceramic composites is based upon the use of glass forming constituents to generate a nominal composition equivalent to Bioglass® grade 45S5 (with a total content of 15 vol. %). In terms of the ceramic component in this composite, it was originally designed to form Si_2N_2O, through the reaction of α-Si_3N_4 and an equimolar amount of SiO_2. The foaming response of this material is demonstrated in Figure 5. It is apparent that the disc-shaped samples essentially double in thickness, while the diameter is largely unchanged, showing a maximum of ~10 % shrinkage (Figure 5(a)). This expansion is also highlighted in terms of the decrease in density, as shown in Figure 5(b).

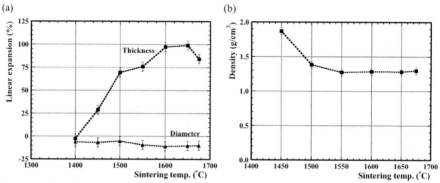

Figure 5. The effects of heat-treatment temperature upon Si_3N_4-Si_2N_2O-Bioglass® composites: (a) dimensional changes and (b) measured immersion density.

The reduction in density from the initial pressed pellet is essentially consistent when conducting heat-treatments at temperatures of 1500°C, and above. The effects of temperature upon the developed foam microstructure are shown in the FE-SEM micrographs presented in Figure 6. The progressive coarsening of the composite foam cell size is readily apparent, with foams prepared at 1500°C and above having a clear and relatively uniform cell formation at each individual temperature. At the highest temperature (1675°C), the effects of varying the heat-treatment time can also be seen in Figure 6(c,d), with coarsening and cell coalescence apparent, especially after 4 hours.

While perhaps not immediately clear from the images in Figure 6, in terms of foam formation it was apparent that foaming is actually initiated at the outer periphery of the *underside* of the samples during the heat-treatment, rather than the top. The foamed segment of the samples then extends along the bottom surface and up the sides of the samples as the temperature is increased. This foaming behavior was unexpected, and a full explanation is not possible at the present time. The edge of the samples will present the highest adjacent surface area, and this likely helps in terms of bubble nucleation as the external walls can deform more easily. The foam cell growth subsequently progresses both towards the centre of the bottom face, and also through the thickness of the discs. As a consequence of this mechanism a slight gradation in pore size is evident through the sample thickness and from the edge to the middle of the discs; this effect is most apparent in Figure 6(b).

As noted earlier, through the use of TGA it was established that the Bioglass® precursors (i.e. carbonates and hydrates) had essentially lost their volatile components following the early ramp and hold at 800°C. It is therefore notable that no foaming was observed below 1300°C. This response indicates that the foaming process is likely initiated through decomposition of either the oxide glass constituents themselves, or potentially through volatilization of Si_3N_4. While neither of these phenomena can be completely ruled out, the former is more likely as weight loss was consistent through the temperature range of 1300 to 1650°C, and only showed a significant increase at 1675°C, where Si_3N_4 dissociation can be expected to become a factor.

(a)   (b)

(c)   (d)

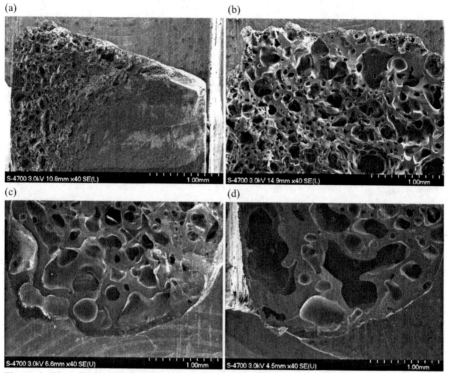

Figure 6. The effects of heat-treatment conditions upon the microstructure of the Si_3N_4-Si_2N_2O-Bioglass® composite foams: (a) 1400°C/2 h (foaming initiation point top left hand side), (b) 1500°C/2 h (foaming initiation point top right hand side), (c) 1675°C/2 (foaming initiation point bottom left hand side) and (d) 1675°C/4 h (foaming initiation point bottom left hand side).

Initially the aim of the present work was to generate a Si_2N_2O-Bioglass® composite, where the orthorhombic Si_2N_2O needles would potentially toughen the relatively brittle Bioglass® material. However, XRD demonstrates that below 1675°C there is no evidence of Si_2N_2O formation, and in fact the excess SiO_2 initially added is retained as β-cristobalite (Figure 7). It is also apparent that the α-Si_3N_4 starting powder is retained to noticeably higher temperature than is observed for typical oxide sintering additives used with Si_3N_4 (e.g. compare with Figure 4(b), where between 40 and 70 vol. % β-Si_3N_4 has been formed by 1650°C). However, for the sample prepared at 1675°C it is clear that two things have happened. Firstly, the β-cristobalite peak has essentially disappeared and secondly, the formation of Si_2N_2O has clearly arisen.

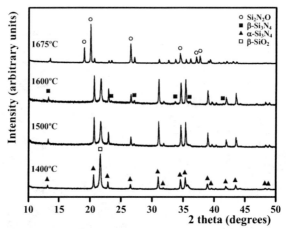

Figure 7. Phase evolution as a function of heat-treatment temperature for the Si₃N₄-Si₂N₂O-Bioglass®
foams; in each case the samples were heat-treated for 2 h.

CONCLUSIONS

In the present work the use of conventional air-atmosphere sintering furnaces for has been demonstrated for the preparation of both porous Si₃N₄ ceramics and novel Si₃N₄-Si₂N₂O-Bioglass® foams. The effects of heating rate have been examined, and it is shown for porous Si₃N₄ ceramics that slower heating rates lead to sample oxidation and increased densification, which is undesired in the current scenario. The porosity in the final sample can also be varied through the incorporation of fugitive filler particles during green processing, such that the fillers are subsequently eliminated through the use of a low temperature oxidation treatment. It has also been demonstrated that both the densification and α- to β-Si₃N₄ phase transformation behavior can be subtly influenced by the choice of rare-earth additive used. An adaptation of the air-atmosphere sintering approach has also allowed the formation of novel, foamed Si₃N₄-Si₂N₂O-Bioglass® composites. The extent of foaming is, in part, dependent upon the temperature (and hence the glass viscosity), and likely arises from partial decomposition of one or more of the glass forming additions. Further work is needed to assess the implications of such behavior, especially from the perspective of bio-activity of the glass.

ACKNOWLEDGEMENTS

The Natural Sciences and Engineering Research Council of Canada (NSERC) are gratefully acknowledged for funding this work through both the Discovery Grants and Undergraduate Student Research Awards programs. The support of the Canada Foundation for Innovation, the Atlantic Innovation Fund, and other partners who helped fund the Facilities for Materials Characterisation, managed by the Dalhousie University Institute for Materials Research, is also gratefully acknowledged, through access to the scanning electron microscopy facilities.

REFERENCES
[1]M.H. Bocanegra-Bernal and B. Matovic, "Mechanical Properties of Silicon Nitride-based Ceramics and its use in Structural Applications at High Temperatures," *Mater. Sci. Eng. A*, **527** [6] 1314-1338 (2010).

[2]Z. Krstic and V.D. Krstic, "Silicon Nitride: The Engineering Material of the Future," *J. Mater. Sci.*, **47** [2] 535-552 (2012).

[3]S. Wada, T. Hattori and K. Yokoyama, "Sintering of Si₃N₄ in Air Atmosphere Furnace," *J. Ceram. Soc. Japan*, **109** [3] 281-283 (2001).

[4]K.P. Plucknett and H.-T. Lin, "Sintering Silicon Nitride Ceramics in Air," *J. Am. Ceram. Soc.*, **88** [12] 3538 (2005).

[5]K.P. Plucknett, "Sintering Behavior and Microstructure Development of Porous Silicon Nitride Ceramics Prepared in an Air Atmosphere Furnace," *Int. J. Appl. Ceram. Tech.*, **6** [6] 702-716 (2009).

[6]K.P. Plucknett, "Processing Factors Involved in Sintering β-Si₃N₄-Based Ceramics in an Air Atmosphere Furnace," *Ceram. Eng. Sci. Proc.*, **30** [8] 45-52 (2009).

[7]P. Chanda, L.B. Garrido, L.A. Genova and K.P. Plucknett, "Porous β-Si₃N₄ Ceramics Prepared with Fugitive Graphite Filler," *Ceram. Eng. Sci. Proc.*, **30** [6] 281-290 (2009).

[8]M. Quinlan, D. Heard, L.B. Garrido, L.A. Genova and K.P. Plucknett, 'Sintering Behaviour and Microstructure Development of Porous Silicon Nitride Ceramics in the Si-RE-O-N Quaternary Systems (RE = La, Nd, Sm, Y, Yb),' *Ceram. Eng. Sci. Proc.*, **28** [9] 41-48 (2007).

COMPARISON OF ELASTIC MODULI OF POROUS CORDIERITE BY FLEXURE AND DYNAMIC TEST METHODS

R. J. Stafford, K. B. Golovin and A. Dickinson, Cummins Inc, Columbus, IN

T. R. Watkins, A. Shyam and E. Lara-Curzio, Oak Ridge National Laboratory, Oak Ridge, TN

ABSTRACT

Previous work[1] showed differences in apparent elastic modulus between mechanical flexure testing and dynamic methods. Flexure tests have been conducted using non-contact optical systems to directly measure deflection for calculation of elastic modulus. Dynamic test methods for elastic modulus measurement were conducted on the same material for comparison. The results show significant difference in the apparent elastic modulus for static flexure versus dynamic methods. The significance of the difference in apparent elastic modulus on thermal stress and the hypotheses for these differences will be discussed.

INTRODUCTION

Young's modulus is a physical property derived from the strain response of a material in reaction to an applied stress (Hooke's Law). For a ceramic, this application of stress typically results in a linear strain response until brittle fracture at low strain values. The assumption of a homogeniety and isotropy for a polycrystalline sample is typically valid. The addition of porosity to the ceramic material does not usually change the fundamental application of linear elasticity, and one may assume that the material behaves according to a rule of mixtures, is homogenous and is isotropic. The derived elastic modulus is then an effective modulus. There are a large number of literature articles on relationships of elastic modulus to porosity in ceramics with a good summary given by Pabst[2]. Many semi-empirical relationships have been developed to model the modulus-porosity relationship with primary success for isotropic materials having porosity of 0.10 to 0.40 volume fraction. These relationships break down when higher porosity (<0.5 volume fraction) is encountered, as in the cordierite diesel particulate filter (DPF) materials. Since predictive equations for these higher porosity materials are lacking, experimental methods are necessary to evaluate the effective modulus.

Methods available to measure elastic modulus include static methods such as tensile, flexure, and nano-indentation and dynamic methods such as resonant ultrasound spectroscopy (RUS), dynamic mechanical analysis (DMA), pulse excitation, and sonic velocity. Static and dynamic methods have been compared for dense alumina, glass, aluminum and steel[3] with good agreement between static and dynamic methods. Extending this methodology to porous ceramics resulted in a finding of significant difference between apparent elastic modulus from static and dynamic methods[1]. Further work on cordierite ceramic was undertaken to determine if the differences between static and dynamic methods are due to inherent differences in material response to stress application or an artifact of the test method application. Understanding the difference is important as elastic modulus is used in many stress models to predict material response and fracture. A significant change in modulus would have a pronounced effect on the predicted stresses and subsequent reliability calculations.

EXPERIMENTAL

Sample filters of cordierite Duratrap AC[*] (nominal size 200 cpsi/8 mil wall[†]) were used to create specimens for testing. Multiple flexure bars were cut from two filters and randomized. The flexure bars were approximately 12 x 25 x 150 mm. Each flexure bar was tested for elastic modulus using a Buzz-o-sonic[‡] tester in accordance with ASTM E1876-09[4] for Out-of-Plane Flexure. Four specimen bars were then sectioned to make multiple two cell x four cell x 60 mm specimens for Dynamic Mechanical Analysis (DMA) testing. These specimens were tested using a Q800 DMA tester[§] with a three point bending fixture operated with a deflection amplitude of 5 μm in a frequency sweep with test points at 0.1, 1, 10 and 100 Hz. The remaining flexure specimens were tested in accordance with ASTM C1674-11[5] using a Sintech 20G universal test machine[**] with a 500 N load cell, fully articulating four point bend fixture with 13 mm rollers, 45 mm loading span and 90 mm support span. The crosshead speed for testing was 0.5 mm/min. The flexure specimens were separated into two groups with the deflection of each group monitored by non-contact methods. The two methods were Vision System (VS) and Digital Image Correlation (DIC). Method 1 was a National Instruments (NI) Vision Camera System[††] which measures deflection by identifying a group of pixels with a camera and then tracking the pixel cluster as the specimen deformed. Method 2 was a Vic-3D 2010 Digital Image Correlation System[‡‡] which measures deflection by using two cameras to track a speckle painted target. The speckle painting is similar to the selected cluster of pixels used by the NI Vision system. The center of the samples was tracked for deflection relative to the same left outer support. The camera systems were synced to the load cell output from the Sintech 20G load frame to develop the load deflection curve for determination of the elastic modulus.

Since the above test methods were developed for solid cross section specimens, the honeycomb structure of the specimens must be accounted for in the elastic modulus and stress calculations. To determine the moment of inertia, I, a minimum of ten measurements of the ceramic wall thickness and cell pitch were taken on each end of every specimen using a Keyence[§§] VHX-500 digital microscope calibrated at the magnification and focal length used for measurements.

The output from the different tests was fundamental response measurements of frequency and/or load and deflection. These measurements were then converted to apparent elastic modulus.

The resonant frequency from pulse excitation was used with the measured dimensions and mass and calculated moment of inertia to determine the apparent elastic modulus by Equation 1.

Equation 1[6]:
$$E = \frac{f^2 m L^3}{12.674 I}$$

where E, f, m and L are the Young's modulus, fundamental frequency, sample mass and length, respectively. For DMA, the apparent elastic modulus was recorded directly as the storage modulus in the results as calculated by Equation 2. Inputs to the program included the moment of inertia and cross section area of the honeycomb specimen. Cross section area was determined using the area calculation function in ImageJ[7].

Equation 2:
$$E' = \frac{\sigma_0}{\varepsilon_0} \cos \delta$$

where E', σ_0, ε_0 and δ are the Storage modulus (apparent Young's modulus), applied stress (corrected for moment of inertia), measured strain and phase angle between applied stress and measured strain, respectively. The apparent elastic modulus for mechanical tests was calculated from the load and deflection of the specimens by Equation 3. The deflection for VS and DIC were taken directly from the measurement output of the camera systems. The deflection measured by the crosshead movement was corrected for machine compliance using a known modulus material as a reference[8].

Equation 3[9]:
$$E = \frac{PL_S^3}{69.8Iy}$$

where P, L_S and y are the applied load, support span length and deflection distance, respectively, for a sample loaded in four point flexure with a moment arm of $L_S/4$.

RESULTS

The calculated apparent elastic modulus values for the dynamic and static test methods are shown in Figure 1. These results include corrections for moment of inertia of the honeycomb structure for all methods and machine compliance for the four point flexure method using crosshead deflection. Machine compliance was calculated using a copper bar with known elastic modulus of 124.5 GPa. The static test method (flexure) shows much lower elastic moduli (~50%) than the dynamic test methods (pulse excitation and DMA). Representative test outputs for each of the methods are shown in Figure 2.

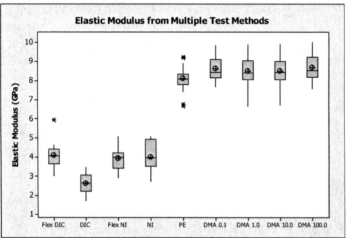

Figure 1 – Apparent Elastic Modulus comparison. Flex DIC and Flex NI are results using crosshead movement to calculate deflection. DIC (digital image correlation) and NI (National Instruments vision system) are results using non-contact measurement of deflection.

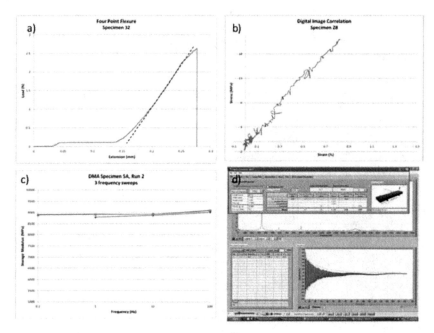

Figure 2 - Representative test output for each method. Outputs are different for each method a) load, b) stress, c) storage modulus, d) frequency.

DISCUSSION

A summary of the calculated elastic modulus data is shown in Table I with the t-test analysis between the different test methods. The comparison shows statistical differences in the sample populations between the static and dynamic test methods. The non-contact Vision System correlates with the two data sets for four point flexure using crosshead deflection. This is an indication that the correction for machine compliance works. The Digital Image Correlation has a lower value than the four point flexure and Vision System, and DIC is found to be significantly different. Review of the DIC set-up showed that the image area was restricted to approximately 65 mm of the support span due to design of the fixture occluding the view of the specimen surface. The reduced length of image area may be introducing an error in the deflection measurement. Further work on DIC is needed where the entire specimen surface is in view of the camera during the test to ensure accurate recording of the specimen deflection.

Comparison of the dynamic methods (pulse excitation and DMA) showed that there is a statistical difference between the methods. Additional comparison of the DMA method over four different frequencies showed no statistical difference between frequencies of 0.1 to 100 Hz.

Table I – Elastic Modulus Summary

Test		Elastic Modulus			t-test	
		Mean (GPa)	St Dev (GPa)	# data points	Comparison	P value*
1	4 point flex A	4.10	0.652	16		
2	Digital Image A	2.63	0.499	14	1 vs 2	0.000
3	4 point flex B	3.94	0.634	14	1 vs 3	0.483
4	Vision System B	4.01	0.724	15	3 vs 4	0.783
5	Pulse Excitation	8.10	0.547	40	5 vs 7	0.001
6	DMA 0.1 Hz	8.62	0.583	98		
7	DMA 1 Hz	8.50	0.634	122	6 vs 7	0.156
8	DMA 10 Hz	8.50	0.646	123	6 vs 8	0.154
9	DMA 100 Hz	8.68	0.610	137	6 vs 9	0.438

* - P value is a statistical measurement for correlation of sample populations. P value < 0.05 indicates that the sample populations being compared do not have equivalent mean values.

Since there is a statistical difference between the static and dynamic methods, there is now a question of why these results are different. The cordierite material has a microstructure of distributed porosity and intentionally formed microcracks which results in the desired low thermal expansion behavior (< 1 ppm/°C between 25 and 1000 °C) for application in exhaust aftertreatment systems. While the distributed porosity would be expected to behave as an additive phase in contribution to the apparent elastic modulus (rule of mixtures) the microcracks present an additional contribution. A hypothesis would be that the microcracks participate in the measurement of the apparent modulus if there is sufficient time for the crack to open or close during active measurement. At high frequency, with very low strain, there would not be sufficient time for crack movement and the microcracks would have no effect on the apparent modulus value and the assumptions of linear elasticity in a homogeneous, isotropic material are valid.

Another hypothesis would be that the mechanical test produces larger strains and the interaction of the microcrack opening/growth with larger strains would cause the load-deflection response to deviate from linear elasticity. In an attempt to test the possible frequency effect, the DMA test was run over a range of frequencies. This showed no difference in apparent modulus. However, the minimum DMA frequency of 0.1 Hz still is much greater than the effective frequency in a flexure test (on the order of 0.0006 Hz) so the frequency difference is too great to draw any conclusions. Tests where the strain application is on the order of 0.01 and 0.001 Hz are needed to provide further insight. Mechanical testing by flexure to reach strain applications in the range of dynamic measurements will require higher precision in the deflection measurement. A non-contact system (Vision System or Digital Image Correlation) with a higher resolution load cell should provide this capability.

In addition to four point flexure testing, there is current work[10] using O-ring compression and biaxial flexure with micro-FEA to determine the apparent elastic modulus of porous honeycomb cordierite ceramic. This work has also confirmed that the apparent mechanical modulus determined from a static (mechanical) test method is significantly lower that the apparent modulus from a dynamic test method. The biaxial flexure results are nearly equivalent to these four point flexure results.

The significance of understanding apparent elastic modulus is in the impact which a change in modulus has on the final application of the material. The specific use of cordierite in diesel aftertreatment systems for particulate filters utilizes elastic modulus as a modeling input to convert temperature profiles into stress profiles. An example of this is shown in Figure 3.

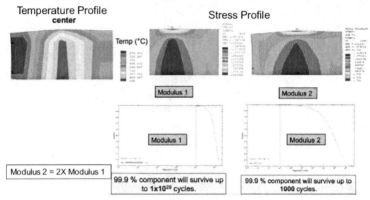

Figure 3 - Thermal Stress and Durability[11]

The stress profile and apparent elastic modulus are also used as inputs to the durability models which have been created as tools to predict the lifetime use for the final aftertreatment components. The effect of a modulus change on the durability prediction is shown in the lower section of Figure 3 where a change of modulus by a factor of two results in a prediction of 17 orders of magnitude reduction in durability. From this result, it can be seen that the use of an accurate and reasonable value for material elastic modulus is required to have a prediction of useful life. In addition, a fundamental rethinking of the meaning of linear elastic constants in the case of porous microcracked ceramics may be necessary.

CONCLUSIONS

Dynamic measurement (resonance) and static measurement (mechanical) produce different values for elastic modulus of porous cordierite ceramic. The elastic modulus from resonance is a measure of the material response at very low strain which is different from the material response in a mechanical test with relatively large strain. The apparent elastic moduli for dynamic versus static test methods in this study are different by a factor of two. This result has significant impact on calculated stress and life in an aftertreatment component.

REFERENCES
1. R. J. Stafford, P. J. Gloeckner, T. R. Watkins, A. Shyam, E. Lara-Curzio, "Elastic Modulus Measurement of Porous Cellular Ceramics Using Multiple Test Methods," presented by R. J. Stafford at the 35th International Conference and Exposition on Advanced Ceramics and Composites, Daytona Beach, FL, January 25, 2011.
2. W. Pabst, E. Gregorova, G. Ticha, "Elasticity of porous ceramics – A critical study of modulus-porosity relations," Journal of the European Ceramic Society, Vol. 26, p 1085-

1097 (2006).
3. M. Radovic, E. Lara-Curzio, L. Reister, "Comparison of Different Experimental Techniques for Determination of Elastic Properties of Solids," Materials Science and Engineering, A368, p. 56 – 70 (2004).
4. ASTM E1876-09, Standard Test Method for Dynamic Young's Modulus, Shear Modulus and Poisson's Ratio by Impulse Excitation of Vibration, ASTM Book of Standards, Vol 3.01, West Conshohocken, NJ, (2011).
5. ASTM C1674-11, Standard Test Method for Flexural Strength of Advanced Ceramics with Engineered Porosity (Honeycomb Cellular Channels) at Ambient Temperature, ASTM Book of Standards, Vol 15.01, West Conshohocken, NJ, (2011).
6. A. H. Church, Mechanical Vibrations, 2nd ed, John Wiley and Sons, Inc. New York 1963, p. 379.
7. ImageJ software, National Institutes of Health freeware, nih.gov.
8. S. R. Kalidindi, A. Abusafieh and E. El-Danaf, "Accurate Characterization of Machine Compliance for Simple Compression Testing", Experimental Mechanics, Vol. 37(2), p 210 – 215, (1997).
9. Popov, E., Engineering Mechanics of Solids, 2nd ed., Prentice Hall, 1998.
10. A. A. Wereszczak, M. J. Lance, E. E. Fox, M. K. Ferber, "Failure Stress and Apparent Elastic Modulus of Diesel Particulate Filter Ceramics", presented at CLEERS Focus Group teleconference, November 17, 2011.
11. T. R. Watkins, A. Shyam, H.T. Lin, E. Lara-Curzio, R. Stafford, T. Yonushonis, "Durability of Diesel Engine Particulate Filters," presented at the DOE 2011 Vehicle Technologies Annual Merit Review and Peer Evaluation Meeting, Washington, D.C., May 11, 2011.

ACKNOWLEDGEMENT

Research at Oak Ridge National Laboratory was sponsored by the U.S. Department of Energy, Assistant Secretary for Energy Efficiency and Renewable Energy, Office of Vehicle Technologies, as part of the Propulsion Materials Program. Some of the equipment and instrumentation utilized during this investigation was acquired and maintained by the Oak Ridge National Laboratory's High Temperature Materials Laboratory User Program, which is sponsored by the U. S. Department of Energy, Office of Energy Efficiency and Renewable Energy, Vehicle Technologies Program.

---

[*] Corning, Inc, Corning, NY
[†] Standard terminology for extruded honeycomb is English units - cells per square inch (cpsi) and mil (0.001 inch)
[‡] BuzzMac International, Glendale, WI
[§] TA Instruments, New Castle, DE
[**] MTS Corporation, Eden Prairie, MN
[††] National Instruments, Austin, TX
[‡‡] Correlated Solutions, Inc., Columbia, SC
[§§] Keyence Corporation, Osaka, Japan

Author Index